Elements of Continuum Mechanics and Thermodynamics

This text is intended to provide a modern and integrated treatment of the foundations and applications of continuum mechanics. There is a significant increase in interest in continuum mechanics because of its relevance to microscale phenomena. In addition to being tailored for advanced undergraduate students and including numerous examples and exercises, this text also features a chapter on continuum thermodynamics, including entropy production in Newtonian viscous fluid flow and thermoelasticity. Computer solutions and examples are emphasized through the use of the symbolic mathematical computing program Mathematica®.

Joanne L. Wegner is a Professor of Mechanical Engineering at the University of Victoria, Canada. Her research interests include wave propation in elastic and viscoelastic materials with applications to the dynamic behavior of polymers and elastomeric materials and earthquake engineering. Professor Wegner has been recognized for her outstanding contributions to the engineering profession by the Canadian Society of Mechanical Engineers. In addition to her archival publications, she has coedited two books on nonlinear waves in solids.

James B. Haddow received the Canadian Congress of Applied Mechanics award in 2003. He has published papers on plasticity, nonlinear elasticity, wave propagation, and thermodynamics (exergy, thermoelasticity).

Elements of Continuum Mechanics and Thermodynamics

Joanne L. Wegner

University of Victoria

James B. Haddow

University of Victoria

CAMBRIDGE
UNIVERSITY PRESS

CAMBRIDGE
UNIVERSITY PRESS

32 Avenue of the Americas, New York NY 10013-2473, USA

Cambridge University Press is part of the University of Cambridge.

It furthers the University's mission by disseminating knowledge in the pursuit of education, learning and research at the highest international levels of excellence.

www.cambridge.org
Information on this title: www.cambridge.org/9781107460140

First published 2009
First paperback edition 2014

A catalogue record for this publication is available from the British Library

Library of Congress Cataloguing in Publication data

Wegner, Joanne L.
Elements of continuum mechanics and thermodynamics / Joanne L. Wegner, James B. Haddow. – 1st ed.
p. cm.
Includes index.
ISBN 978-0-521-86632-3 (hardback)
1. Continuum mechanics – Textbooks. 2. Thermodynamics – Textbooks. I. Haddow, James B. II. Title

QA808.2.W44 2009
531–dc22 2008035724

ISBN 978-0-521-86632-3 Hardback
ISBN 978-1-107-46014-0 Paperback

Contents

Preface *Page* ix

1. **Cartesian Tensor Analysis** 1

1.1 Introduction 1
1.2 Rectangular Cartesian Coordinate Systems 3
1.3 Suffix and Symbolic Notation 4
1.4 Orthogonal Transformations 6
1.5 Concept of a Second-Order Tensor 11
1.6 Tensor Algebra 19
1.7 Invariants of Second-Order Tensors 20
1.8 Cayley-Hamilton Theorem 24
1.9 Higher-Order Tensors 25
1.10 Determinants 28
1.11 Polar Decomposition Theorem 32
1.12 Tensor Fields 38
1.13 Integral Theorems 44
EXERCISES 46
SUGGESTED READING 48

2. **Kinematics and Continuity Equation** 50

2.1 Description of Motion 50
2.2 Spatial and Referential Descriptions 53
2.3 Material Surface 54
2.4 Jacobian of Transformation and Deformation Gradient $\boldsymbol{F}$ 56
2.5 Reynolds Transport Theorem 59
2.6 Continuity Equation 63

2.7 Velocity Gradient 64
2.8 Rate of Deformation and Spin Tensors 65
2.9 Polar Decomposition of ***F*** 70
2.10 Further Decomposition of ***F*** 74
2.11 Strain 75
2.12 Infinitesimal Strain 78
EXERCISES 81

3. **Stress** 84
3.1 Contact Forces and Body Forces 84
3.2 Cauchy or True Stress 87
3.3 Spatial Form of Equations of Motion 90
3.4 Principal Stresses and Maximum Shearing Stress 91
3.5 Pure Shear Stress and Decomposition of the Stress Tensor 94
3.6 Octahedral Shearing Stress 95
3.7 Nominal Stress Tensor 98
3.8 Second Piola-Kirchhoff Stress Tensor 100
3.9 Other Stress Tensors 101
EXERCISES 102
REFERENCES 104

4. **Work, Energy, and Entropy Considerations** 105
4.1 Stress Power 105
4.2 Principle of Virtual Work 110
4.3 Energy Equation and Entropy Inequality 114
EXERCISES 119
REFERENCES 120

5. **Material Models and Constitutive Equations** 121
5.1 Introduction 121
5.2 Rigid Bodies 121
5.3 Ideal Inviscid Fluid 123
5.4 Incompressible Inviscid Fluid 124
5.5 Newtonian Viscous Fluid 125
5.6 Classical Elasticity 126
5.7 Linear Thermoelasticity 130
5.8 Determinism, Local Action, and Material Frame Indifference 138

EXERCISES 147
REFERENCES 149

6. **Finite Deformation of an Elastic Solid** 150
6.1 Introduction 150
6.2 Cauchy Elasticity 150
6.3 Hyperelasticity 155
6.4 Incompressible Hyperelastic Solid 160
6.5 Alternative Formulation 162
EXERCISES 166
REFERENCES 167

7. **Some Problems of Finite Elastic Deformation** 168
7.1 Introduction 168
7.2 Strain Energy Functions and Stress-Strain Relations 169
7.3 Simple Shear of a Rectangular Block 170
7.4 Simple Tension 172
7.5 Extension and Torsion of an Incompressible Cylindrical Bar 175
7.6 Spherically Symmetric Expansion of a Thick-Walled Shell 180
7.7 Eversion of a Cylindrical Tube 185
7.8 Pure Bending of an Hyperelastic Plate 189
7.9 Combined Telescopic and Torsional Shear 195
EXERCISES 200
REFERENCES 201

8. **Finite Deformation Thermoelasticity** 202
8.1 Principle of Local State and Thermodynamic Potentials 202
8.2 Basic Relations for Finite Deformation Thermoelasticity 202
8.3 Ideal Rubber-Like Materials 209
8.4 Isentropic Simple Tension of Ideal Rubber 214
8.5 Isentropic Simple Tension of Compressible Rubber 215
EXERCISES 218
REFERENCES 219

9. **Dissipative Media** 220
9.1 Newtonian Viscous Fluids 220

9.2 A Non-Newtonian Viscous Fluid 223
9.3 Linear Viscoelastic Medium 224
9.4 Viscoelastic Dissipation 237
9.5 Some Thermodynamic Considerations in Linear Thermoelasticity 240
9.6 Simple Shear Problem 244
EXERCISES 246
REFERENCES 248

APPENDIX 1: Orthogonal Curvilinear Coordinate Systems 249
A1.1 Introduction 249
A1.2 Curvilinear Coordinates 249
A1.3 Orthogonality 252
A1.4 Cylindrical and Spherical Polar Coordinate Systems 253
A1.5 The ∇ Operator 256
A1.6 The ∇^2 Operator 261

APPENDIX 2: Physical Components of the Deformation Gradient Tensor 263

APPENDIX 3: Legendre Transformation 266

APPENDIX 4: Linear Vector Spaces 269

Index 272

Preface

This book is based on notes prepared for a senior undergraduate or beginning graduate course that we taught at the universities of Alberta and Victoria. It is primarily intended for use by students of mechanical and civil engineering, but it may be of interest to others. The mathematical background required for the topics covered in the book is modest and should be familiar to senior undergraduate engineering students. In particular it is assumed that a reader has a good knowledge of classical vector mechanics and linear algebra. Also, a background of the classical thermodynamics usually taught in undergraduate engineering courses is desirable. One motivation for the book is to present an introduction to continuum mechanics that requires no background in certain areas of advanced mathematics such as functional analysis and general tensor analysis. The treatment of continuum mechanics is based on Cartesian tensor analysis, but orthogonal curvilinear coordinates and corresponding physical coordinates are considered in appendices.

A list of books that consider tensor analysis and applications is given at the end of chapter 1. Several of the books are out of print but may be useful to students if they can be obtained from libraries. Mathematica is used for symbolic manipulation, numerical computation, and graphs where appropriate, and its use is encouraged.

Chapter 1 is a detailed introduction to Cartesian tensor analysis. It differs from some other treatments of the topic, for example, the early texts by Jeffreys [1] and Temple [2], by emphasizing both symbolic and suffix notation for first-(vectors) and second-order tensors. Many tensor relations are given in both symbolic and Cartesian suffix notation. The introduction of general tensor analysis involving contravariant and covariant components of tensors and Christoffel symbols is avoided. However, two appendices are included, which introduce the concept of physical components, of first- and second-order tensors, with respect to orthogonal curvilinear coordinates.

The kinematics of a deformable medium is the subject of chapter 2, along with the continuity equation and Reynold's transport theorem. Cauchy, nominal (transpose of Piola-Kirchhoff I), Piola-Kirchhoff II, and Biot stress tensors are introduced in chapter 3, along with equations of motion. In chapter 4 the concept of stress power is related to the first and second laws of thermodynamics. Continuum thermodynamics involves irreversible processes, and what is known as the thermodynamics of irreversible processes (TIP) is assumed. Several classical models for continuous media are discussed in chapter 5, and conditions that should be satisfied for the admissibility of constitutive relations are discussed. Thermodynamic considerations also appear in chapter 5, including a discussion of the constitutive relations for linear thermoelasticity. Nonlinear isotropic elasticity is the subject of chapters 6 and 7. The treatment is for isothermal static deformation, and one purpose is to illustrate the application of certain principles introduced in earlier chapters. Solutions to several boundary value problems are given in chapter 7. Chapter 8 considers nonlinear thermoelasticity. The final chapter (9) is concerned with dissipative materials. A brief discussion of viscous fluids is given, in addition to a discussion of certain aspects of linear viscoelasticity. An elementary example of an internal variable is introduced in chapter 9 in connection with the three-parameter standard model of linear viscoelasticity.

1 Cartesian Tensor Analysis

1.1 Introduction

In this chapter we present an elementary introduction to Cartesian tensor analysis in a three-dimensional Euclidean point space or a two-dimensional subspace. A Euclidean point space is the space of position vectors of points. The term vector is used in the sense of classical vector analysis, and scalars and polar vectors are zeroth- and first-order tensors, respectively. The distinction between polar and axial vectors is discussed later in this chapter. A scalar is a single quantity that possesses magnitude and does not depend on any particular coordinate system, and a vector is a quantity that possesses both magnitude and direction and has components, with respect to a particular coordinate system, which transform in a definite manner under change of coordinate system. Also vectors obey the parallelogram law of addition. There are quantities that possess both magnitude and direction but are not vectors, for example, the angle of finite rotation of a rigid body about a fixed axis.

A second-order tensor can be defined as a linear operator that operates on a vector to give another vector. That is, when a second-order tensor operates on a vector, another vector, in the same Euclidean space, is generated, and this operation can be illustrated by matrix multiplication. The components of a vector and a second-order tensor, referred to the same rectangular Cartesian coordinate system, in a three-dimensional Euclidean space, can be expressed as a (3×1) matrix and a (3×3) matrix, respectively. When a second-order tensor operates on a vector, the components of the resulting vector are given by the matrix product of the (3×3) matrix of components of the second-order tensor and the matrix of the (3×1)

components of the original vector. These components are with respect to a rectangular Cartesian coordinate system, hence the term Cartesian tensor analysis. Examples from classical mechanics and stress analysis are as follows. The angular momentum vector, $\boldsymbol{h}$, of a rigid body about its mass center is given by $\boldsymbol{h} = \boldsymbol{J}\boldsymbol{\omega}$ where $\boldsymbol{J}$ is the inertia tensor of the body about its mass center and $\boldsymbol{\omega}$ is the angular velocity vector. In this equation the components of the vectors $\boldsymbol{h}$ and $\boldsymbol{\omega}$ can be represented by (3×1) matrices and the tensor $\boldsymbol{J}$ by a (3×3) matrix with matrix mutiplication implied. A further example is the relation $\boldsymbol{t} = \boldsymbol{\sigma}\boldsymbol{n}$, between the stress vector $\boldsymbol{t}$ acting on a material area element and the unit normal $\boldsymbol{n}$ to the element, where $\boldsymbol{\sigma}$ is the Cauchy stress tensor. The relations $\boldsymbol{h} = \boldsymbol{J}\boldsymbol{\omega}$ and $\boldsymbol{t} = \boldsymbol{\sigma}\boldsymbol{n}$ are examples of coordinate-free symbolic notation, and the corresponding matrix relations refer to a particular coodinate system.

We will meet further examples of the operator properties of second-order tensors in the study of continuum mechanics and thermodynamics.

Tensors of order greater than two can be regarded as operators operating on lower-order tensors. Components of tensors of order greater than two cannot be expressed in matrix form.

It is very important to note that physical laws are independent of any particular coordinate system. Consequently, equations describing physical laws, when referred to a particular coordinate system, must transform in definite manner under transformation of coordinate systems. This leads to the concept of a tensor, that is, a quantity that does not depend on the choice of coordinate system. The simplest tensor is a scalar, a zeroth-order tensor. A scalar is represented by a single component that is invariant under coordinate transformation. Examples of scalars are the density of a material and temperature.

Higher-order tensors have components relative to various coordinate systems, and these components transform in a definite way under transformation of coordinate systems. The velocity $\boldsymbol{v}$ of a particle is an example of a first-order tensor; henceforth we denote vectors, in symbolic notation, by lowercase bold letters. We can express $\boldsymbol{v}$ by its components relative to any convenient coordinate system, but since $\boldsymbol{v}$ has no preferential relationship to any particular coordinate system, there must be a definite relationship between components of $\boldsymbol{v}$ in different coordinate systems. Intuitively, a vector may be regarded as a directed line segment, in a three-dimensional Euclidean point space E_3, and the set of directed line segments in E_3, of classical vectors, is a vector space V_3. That is, a classical vector is the difference of two points in E_3. A vector, according to this concept, is a first-order tensor. A discussion of linear vector spaces is given in Appendix 4.

There are many physical laws for which a second-order tensor is an operator associating one vector with another. Remember that physical laws must be independent of a coordinate system; it is precisely this independence that motivates us to study tensors.

1.2 Rectangular Cartesian Coordinate Systems

The simplest type of coordinate system is a rectangular Cartesian system, and this system is particularly useful for developing most of the theory to be presented in this text.

A rectangular Cartesian coordinate system consists of an orthonormal basis of unit vectors $(\mathbf{e}_1, \mathbf{e}_2, \mathbf{e}_3)$ and a point 0 which is the origin. Right-handed Cartesian coordinate systems are considered, and the axes in the $(\mathbf{e}_1, \mathbf{e}_2, \mathbf{e}_3)$ directions are denoted by $0x_1$, $0x_2$, and $0x_3$, respectively, rather than the more usual $0x$, $0y$, and $0z$. A right-handed system is such that a 90° right-handed screw rotation along the $0x_1$ direction rotates $0x_2$ to $0x_3$, similarly a right-handed rotation about $0x_2$ rotates $0x_3$ to $0x_1$, and a right-handed rotation about $0x_3$ rotates $0x_1$ to $0x_2$.

A right-handed system is shown in Figure 1.1. A point, $x \in \boldsymbol{E}_3$, is given in terms of its coordinates (x_1, x_2, x_3) with respect to the coordinate system $0x_1x_2x_3$ by

$$\boldsymbol{x} = x_1\mathbf{e}_1 + x_2\mathbf{e}_2 + x_3\mathbf{e}_3,$$

which is a bound vector or position vector.

If points $\boldsymbol{x}, \boldsymbol{y} \in \boldsymbol{E}_3$, $\boldsymbol{u} = \boldsymbol{x} - \boldsymbol{y}$ is a vector, that is, $\boldsymbol{u} \in \mathrm{V}_3$. The vector $\boldsymbol{u}$ is given in terms of its components (u_1, u_2, u_3), with respect to the rectangular

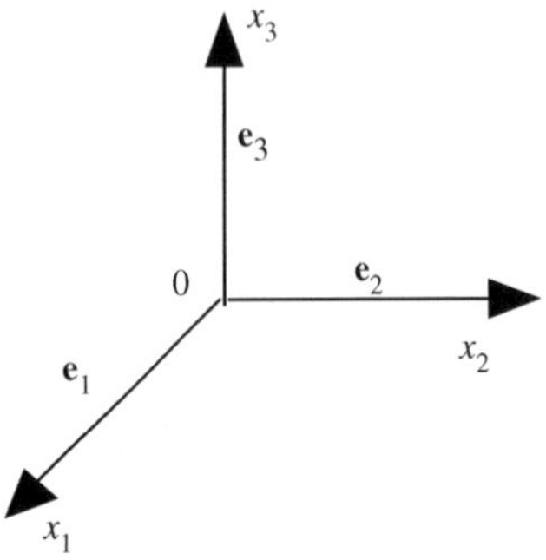

Figure 1.1. Right-handed rectangular Cartesian coordinate system.

coordinate system, $0x_1x_2x_3$ by

$$\boldsymbol{u} = u_1\mathbf{e}_1 + u_2\mathbf{e}_2 + u_3\mathbf{e}_3.$$

Henceforth in this chapter when the term coordinate system is used, a rectangular Cartesian system is understood. When the components of vectors and higher-order tensors are given with respect to a rectangular Cartesian coordinate system, the theory is known as Cartesian tensor analysis.

1.3 Suffix and Symbolic Notation

Suffixes are used to denote components of tensors, of order greater than zero, referred to a particular rectangular Cartesian coordinate system. Tensor equations can be expressed in terms of these components; this is known as suffix notation. Since a tensor is independent of any coordinate system but can be represented by its components referred to a particular coordinate system, components of a tensor must transform in a definite manner under transformation of coordinate systems. This is easily seen for a vector. In tensor analysis, involving oblique Cartesian or curvilinear coordinate systems, there is a distinction between what are called contravariant and covariant components of tensors but this distinction disappears when rectangular Cartesian coordinates are considered exclusively.

Bold lower- and uppercase letters are used for the symbolic representation of vectors and second-order tensors, respectively. Suffix notation is used to specify the components of tensors, and the convention that a lowercase letter suffix takes the values 1, 2, and 3 for three-dimensional and 1 and 2 for two-dimensional Euclidean spaces, unless otherwise indicated, is adopted. The number of distinct suffixes required is equal to the order of the tensor. An example is the suffix representation of a vector $\boldsymbol{u}$, with components (u_1, u_2, u_3) or u_i, $i \in \{1, 2, 3\}$. The vector is then given by

$$\boldsymbol{u} = \sum_{i=1}^{3} u_i \mathbf{e}_i. \tag{1.1}$$

It is convenient to use a summation convention for repeated letter suffixes. According to this convention, if a letter suffix occurs twice in the same term, a summation over the repeated suffix from 1 to 3 is implied without a summation sign, unless otherwise indicated. For example, equation (1.1) can be written as

$$\boldsymbol{u} = u_i \mathbf{e}_{\mathrm{i}} = u_1 \mathbf{e}_1 + u_2 \mathbf{e}_2 + u_3 \mathbf{e}_3 \tag{1.2}$$

without the summation sign. The sum of two vectors is commutative and is given by

$$\boldsymbol{u} + \boldsymbol{v} = \boldsymbol{v} + \boldsymbol{u} = (u_i + v_i)\mathbf{e}_i,$$

which is consistent with the parallelogram rule. A further example of the summation convention is the scalar or inner product of two vectors,

$$\boldsymbol{u} \cdot \boldsymbol{v} = u_i v_i = u_1 v_1 + u_2 v_2 + u_3 v_3. \tag{1.3}$$

Repeated suffixes are often called dummy suffixes since any letter that does not appear elsewhere in the expression may be used, for example,

$$u_i v_i = u_j v_j.$$

Equation (1.3) indicates that the scalar product obeys the commutative law of algebra, that is,

$$\boldsymbol{u} \cdot \boldsymbol{v} = \boldsymbol{v} \cdot \boldsymbol{u}.$$

The magnitude $|\boldsymbol{u}|$ of a vector $\boldsymbol{u}$ is given by

$$|\boldsymbol{u}| = \sqrt{\boldsymbol{u} \cdot \boldsymbol{u}} = \sqrt{u_i u_i}.$$

Other examples of the use of suffix notation and the summation convention are

$$\begin{aligned} C_{ii} &= C_{11} + C_{22} + C_{33} \\ C_{ij} b_j &= C_{i1} b_1 + C_{i2} b_2 + C_{i3} b_3. \end{aligned}$$

A suffix that appears once in a term is known as a free suffix and is understood to take in turn the values 1, 2, 3 unless otherwise indicated. If a free suffix appears in any term of an equation or expression, it must appear in all the terms.

1.4 Orthogonal Transformations

The scalar products of orthogonal unit base vectors are given by

$$\mathbf{e}_i \cdot \mathbf{e}_j = \delta_{ij}, \tag{1.4}$$

where δ_{ij} is known as the Kronecker delta and is defined as

$$\delta_{ij} = \begin{cases} 1 \text{ for } i = j \\ 0 \text{ for } i \neq j \end{cases}. \tag{1.5}$$

The base vectors $\mathbf{e}_i$ are orthonormal, that is, of unit magnitude and mutually perpendicular to each other. The Kronecker delta is sometimes called the substitution operator because

$$u_j \delta_{ij} = u_1 \delta_{i1} + u_2 \delta_{i2} + u_3 \delta_{i3} = u_i. \tag{1.6}$$

Consider a right-handed rectangular Cartesian coordinate system $0x'_i$ with the same origin as $0x_i$ as indicated in Figure 1.2. Henceforth, primed quantities are referred to coordinate system $0x'_i$.

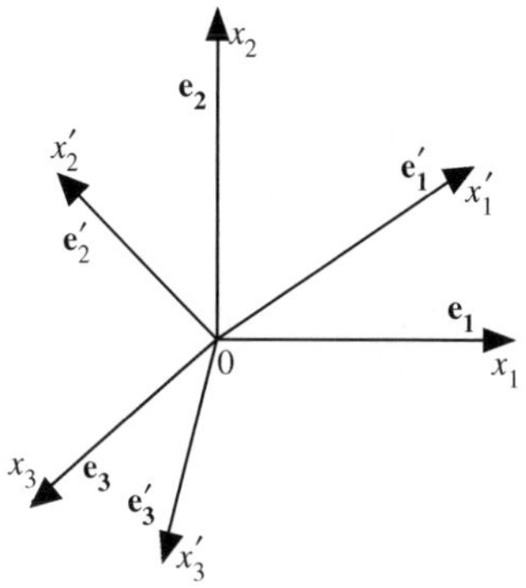

Figure 1.2. Change of axes.

The coordinates of a point P are x_i with respect to $0x_i$ and x_i' with respect to $0x_i$. Consequently,

$$x_i \mathbf{e}_i = x_j' \mathbf{e}_j', \tag{1.7}$$

where the $\mathbf{e}_i'$ are the unit base vectors for the system $0x_i$. Forming the inner product of each side of equation (1.7) with $\mathbf{e}_k'$ and using equation (1.4) and the substitution operator property equation (1.6) gives

$$x_k' = a_{ki} x_i, \tag{1.8}$$

where

$$a_{ki} = \mathbf{e}_k' \cdot \mathbf{e}_i = \cos(x_k' 0 x_i). \tag{1.9}$$

Similarly

$$x_i = a_{ki} x_k'. \tag{1.10}$$

It is evident that the direction of each axis $0x_k'$ can be specified by giving its direction cosines $a_{ki} = \mathbf{e}_k' \cdot \mathbf{e}_i = \cos(x_k' 0 x_i)$ referred to the original axes $0x_i$. The direction cosines, $a_{ki} = \mathbf{e}_k' \cdot \mathbf{e}_i$, defining this change of axes are tabulated in Table 1.1.

The matrix $[\boldsymbol{a}]$ with elements a_{ij} is known as the transformation matrix; it is not a tensor.

Table 1.1. *Direction cosines for rotation of axes*

	e_1'	e_2'	e_3'
e_1	a_{11}	a_{21}	a_{31}
e_2	a_{12}	a_{22}	a_{32}
e_3	a_{13}	a_{23}	a_{33}

It follows from equations (1.8) and (1.10) that

$$a_{ki} = \frac{\partial x'_k}{\partial x_i} = \frac{\partial x_i}{\partial x'_k}, \tag{1.11}$$

and from equation (1.7) that

$$\mathbf{e}_\mathrm{i} \frac{\partial x_i}{\partial x'_k} = \mathbf{e}_\mathrm{j}' \frac{\partial x'_j}{\partial x'_k} = \mathbf{e}'_k, \tag{1.12}$$

since $\partial \boldsymbol{x}_j / \partial \boldsymbol{x}_k = \partial_{jk}$, and from equations (1.11) and (1.12) that

$$\mathbf{e}'_k = a_{ki}\mathbf{e}_i, \tag{1.13}$$

and

$$\mathbf{e}_i = a_{ki}\mathbf{e}'_k. \tag{1.14}$$

Equations (1.13) and (1.14) are the transformation rules for base vectors. The nine elements of a_{ij} are not all independent, and in general,

$$a_{ki} \neq a_{ik}.$$

A relation similar to equations (1.8) and (1.10),

$$u'_k = a_{ki}u_i, \text{ and } u_i = a_{ki}u'_k \tag{1.15}$$

is obtained for a vector $\boldsymbol{u}$ since $u_i\mathbf{e}_i = u'_k\mathbf{e}'_k$, which is similar to equation (1.7) except that the u_i are the components of a vector and the x_i are coordinates of a point.

The magnitude $|\boldsymbol{u}| = (u_i u_i)^{1/2}$ of the vector $\boldsymbol{u}$ is independent of the orientation of the coordinate system, that is, it is a scalar invariant; consequently,

$$u_i u_i = u'_k u'_k. \tag{1.16}$$

Eliminating u_i from equation (1.15) gives

$$u'_k = a_{ki} a_{ji} u'_j,$$

and since $u'_k = \delta_{kj} u'_j$,

$$a_{ki} a_{ji} = \delta_{kj}. \tag{1.17}$$

Similarly, eliminating u_k from equation (1.15) gives

$$a_{ik} a_{jk} = \delta_{ij}. \tag{1.18}$$

It follows from equation (1.17) or (1.18) that

$$\{\det[a_{ij}]\}^2 = 1, \tag{1.19}$$

where $\det[a_{ij}]$ denotes the determinant of a_{ij}. A detailed discussion of determinants is given in section 10 of this chapter. The negative root of equation (1.19) is not considered unless the transformation of axes involves a change of orientation since, for the identity transformation $x_i = x'_i$, $a_{ik} = \delta_{ik}$ and $\det[\delta_{ik}] = 1$. Consequently, $\det[a_{ik}] = 1$, provided the transformations involve only right-handed systems (or left-handed systems).

The transformations (1.8), (1.10), and (1.15) subject to equation (1.17) or (1.18) are known as orthogonal transformations. Three quantities u_i are the components of a vector if, under orthogonal transformation, they transform according to equation (1.15). This may be taken as a definition of a vector. According to this definition, equations (1.8) and (1.10) imply that the representation $\boldsymbol{x}$ of a point is a bound vector since its origin coincides with the origin of the coordinate system.

If the transformation rule (1.10) holds for coordinate transformations from right-handed systems to left-handed systems (or vice versa), the vector is known as a polar vector. There are scalars and vectors known as pseudo scalars and pseudo or axial vectors; there have transformation rules that involve a change in sign when the coordinate transformation is from a right-handed system to a left-handed system (or vice versa), that is, when $\det[a_{ij}] = -1$. The transformation rule for a pseudo scalar is

$$\phi' = \det[a_{ij}]\phi, \tag{1.20}$$

and for a pseudo vector

$$u_i' = \det[a_{ij}]a_{ij}u_j. \tag{1.21}$$

A pseudo scalar is not a true scalar if a scalar is defined as a single quantity invariant under all coordinate transformations. An example of a pseudo vector is the vector product $\boldsymbol{u} \times \boldsymbol{v}$ of two polar vectors $\boldsymbol{u}$ and $\boldsymbol{v}$. A discussion of the vector product is given in section 9 of this chapter. The moment of a force about a point and the angular momentum of a particle about a point are pseudo vectors. The scalar product of a polar vector and a pseudo vector is a pseudo scalar; an example is the moment of a force about a line. The distinction between pseudo vectors and scalars and polar vectors and true scalars disappears when only right- (or left-) handed coordinate systems are considered. For the development of continuum mechanics presented in this book, only right-handed systems are used.

EXAMPLE PROBLEM 1.1. Show that a rotation through angle π about an axis in the direction of the unit vector $\boldsymbol{n}$ has the transformation matrix

$$a_{ij} = -\delta_{ij} + 2n_in_j, \det[a_{ij}] = 1.$$

SOLUTION. Referring to Figure 1.3, the position vector of point A has components x_i and point B has position vector with components x_i'.

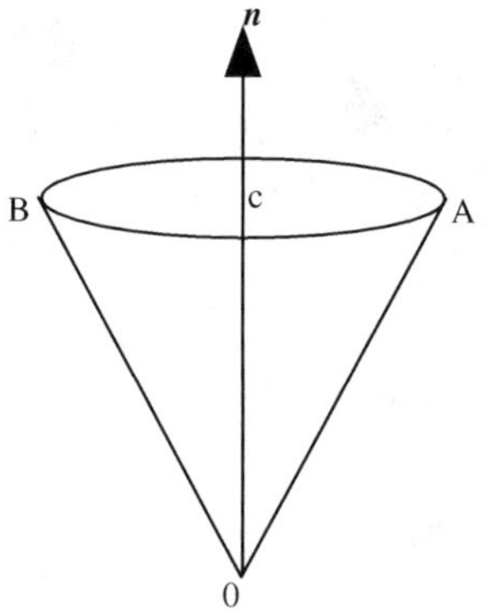

Figure 1.3. Rotation about axis with direction $\boldsymbol{n}$.

Since $AC = CB$, it follows from elementary vector analysis that

$$x_i' = x_i + 2(x_s n_s n_i - x_i) = -\delta_{ij}x_j + 2(\delta_{sj}n_s n_i)x_j = -\delta_{ij}x_j + 2(n_j n_i)x_j$$

so that

$$a_{ij} = -\delta_{ij} + 2n_i n_j.$$

1.5 Concept of a Second-Order Tensor

The tensor or indefinite product, sometimes known as the dyadic or tensor product of two vectors $\boldsymbol{u}, \boldsymbol{v} \in V_3$ is denoted by $\boldsymbol{u} \otimes \boldsymbol{v}$.

This product has no obvious geometrical interpretation, unlike the scalar or vector products, but it is a linear operator that satisfies the relation

$$(\boldsymbol{u} \otimes \boldsymbol{v})\boldsymbol{c} = \boldsymbol{u}(\boldsymbol{v} \cdot \boldsymbol{c}), \tag{1.22}$$

where $\boldsymbol{c}$ is any vector. The tensor product has nine components obtained as follows:

$$(\boldsymbol{u} \otimes \boldsymbol{v})_{ij} = u_i v_j, \tag{1.23}$$

or, in matrix notation,

$$[\boldsymbol{u} \otimes \boldsymbol{v}]_{ij} = \begin{bmatrix} u_1 \\ u_2 \\ u_3 \end{bmatrix} [\boldsymbol{v}_1 \boldsymbol{v}_2 \boldsymbol{v}_3] = \begin{bmatrix} u_1v_1 & u_1v_2 & u_1v_3 \\ u_2v_1 & u_2v_2 & u_2v_3 \\ u_3v_1 & u_3v_2 & u_3v_3 \end{bmatrix}, \tag{1.24}$$

and can be expressed as follows in terms of the basis vectors, $\mathbf{e}_{\mathrm{i}} \otimes \mathbf{e}_{\mathrm{j}}$:

$$\boldsymbol{u} \otimes \boldsymbol{v} = u_i v_j \mathbf{e}_i \otimes \mathbf{e}_j. \tag{1.25}$$

The basis vectors $\mathbf{e}_i \otimes \mathbf{e}_j$ are not vectors in the sense of elementary vector analysis but belong to a nine-dimensional vector space.

A second-order tensor $\boldsymbol{S}$ is a linear transformation from V_3 into V_3 that assigns to each vector $\boldsymbol{t}$ a vector

$$\boldsymbol{s} = \boldsymbol{St}. \tag{1.26}$$

It follows that the tensor product of two vectors is an example of a second-order tensor.

Equation (1.26) is in symbolic form, that is, it has no preferential relationship to any coordinate system, and it may be interpreted in component form by the following matrix equation:

$$\begin{bmatrix} s_1 \\ s_2 \\ s_3 \end{bmatrix} = \begin{bmatrix} S_{11} & S_{12} & S_{13} \\ S_{21} & S_{22} & S_{23} \\ S_{31} & S_{32} & S_{33} \end{bmatrix} \begin{bmatrix} t_1 \\ t_2 \\ t_3 \end{bmatrix}, \tag{1.27}$$

where the square matrix on the right-hand side of equation (1.27) is the matrix of the nine components S_{ij}, of $\boldsymbol{S}$ referred to the rectangular Cartesian coordinate axes $0\boldsymbol{x}_{\mathrm{i}}$. Equation (1.26) or (1.27) can be written in suffix notation as

$$s_i = S_{ij} t_j, \tag{1.28}$$

which, unlike equation (1.26), has a preferential relationship to a particular coordinate system. At this point, it is useful to make a clear distinction between a column matrix and a vector and between a square matrix and a second-order tensor. The components of a vector or second-order tensor may be put in matrix form as in equation (1.27), but a column or square matrix does not necessarily represent the components of a vector or tensor.

The tensor $\boldsymbol{S}$ can be expressed in terms of the basis vectors in the form

$$\boldsymbol{S} = S_{ij}\mathbf{e}_i \otimes \mathbf{e}_j, \tag{1.29}$$

and this shows that a second-order tensor is the weighted sum of tensor products of the base vectors. The left-hand side of equation (1.29) is in coordinate free symbolic form; however, the right-hand side involves the components S_{ij}, which are with respect to a particular coordinate system. Since a tensor is invariant under coordinate transformation,

$$S_{ij}\mathbf{e}_i \otimes \mathbf{e}_j = S'_{kl}\mathbf{e}'_k \otimes \mathbf{e}'_l,$$

and with the use of equation (1.14) the transformation rules

$$S'_{kl} = a_{ki}a_{lj}S_{ij}, \tag{1.30}$$

and

$$S_{ij} = a_{ki}a_{lj}S'_{kl} \tag{1.31}$$

for the componts of $\boldsymbol{S}$ are obtained. It follows from equation (1.29) that the Cartesian components of $\boldsymbol{S}$ are given by

$$S_{ij} = \mathbf{e}_i \cdot \boldsymbol{S}\mathbf{e}_j, \tag{1.32}$$

that is, S_{ij} is the scalar product of the base vector $\mathbf{e}_i$ with the vector $\boldsymbol{S}\mathbf{e}_j$.

It is convenient at this stage to introduce the transpose of a vector and second-order tensor. The components of the transpose $\boldsymbol{u}^T$ of a vector $\boldsymbol{u}$ form a row matrix (u_1, u_2, u_3). The transpose $\boldsymbol{S}^T$ of a second-order tensor $\boldsymbol{S}$ has components $S^T_{ij} = S_{ji}$ so that

$$\boldsymbol{S}^T = S_{ji}\mathbf{e}_i \otimes \mathbf{e}_j.$$

It may be shown that, for any two vectors $\boldsymbol{u}$, $\boldsymbol{v} \in V_3$,

$$\boldsymbol{u} \cdot \boldsymbol{S}^T \boldsymbol{v} = \boldsymbol{v} \cdot (\boldsymbol{S}\boldsymbol{u}).$$

Transposition also applies in general to square matrices as well as the component matrices of second-order tensors.

A tensor $\boldsymbol{S}$ is symmetric if $\boldsymbol{S} = \boldsymbol{S}^T$, that is, $S_{ij} = S_{ji}$. Second-order symmetric tensors in E_3 have six independent components. A second-order tensor $\boldsymbol{T}$ is antisymmetric if $\boldsymbol{T} = -\boldsymbol{T}^T$, that is, $T_{ij} = -T_{ji}$. Second-order antisymmetric tensors in E_3 have three independent components, and diagonal elements of the component matrix are zero. A second-order tensor $\boldsymbol{C}$ can be decomposed uniquely into the sum of a symmetric part and an antisymmetric part as follows:

$$\boldsymbol{C} = \boldsymbol{C}^{(s)} + \boldsymbol{C}^{(a)},$$

where

$$\boldsymbol{C}^{(s)} = \frac{1}{2}\left(\boldsymbol{C} + \boldsymbol{C}^T\right)$$

is the symmetric part and

$$\boldsymbol{C}^{(a)} = \frac{1}{2}\left(\boldsymbol{C} - \boldsymbol{C}^T\right)$$

is the antisymmetric part.

EXAMPLE PROBLEM 1.2. Decompose the tensor $\boldsymbol{C}$ into its symmetric and antisymmetric parts, where the components of $\boldsymbol{C}$ are given by

$$[C_{ij}] = \begin{bmatrix} 1 & 2 & 3 \\ -4 & 2 & 3 \\ 5 & 2 & 1 \end{bmatrix}.$$

SOLUTION. The components of $\boldsymbol{C}^T$ are given by

$$\left[C_{ij}^T\right] = \begin{bmatrix} 1 & -4 & 5 \\ 2 & 2 & 2 \\ 3 & 3 & 1 \end{bmatrix}.$$

Let the tensors $\boldsymbol{C}^{(s)}$ and $\boldsymbol{C}^{(a)}$ denote the symmetric and antisymmetric parts of $\boldsymbol{C}$, respectively; then

$$\left[\boldsymbol{C}^{(s)}\right] = \frac{1}{2}\left[\boldsymbol{C} + \boldsymbol{C}^T\right] = \frac{1}{2}\begin{bmatrix} 2 & -2 & 8 \\ -2 & 4 & 5 \\ 8 & 5 & 2 \end{bmatrix} = \begin{bmatrix} 1 & -1 & 4 \\ -1 & 2 & 2.5 \\ 4 & 2.5 & 1 \end{bmatrix}$$

and

$$\left[\boldsymbol{C}^{(a)}\right] = \frac{1}{2}\left[\boldsymbol{C} - \boldsymbol{C}^T\right] = \frac{1}{2}\begin{bmatrix} 0 & 6 & -2 \\ -6 & 0 & 1 \\ 2 & -1 & 0 \end{bmatrix} = \begin{bmatrix} 0 & 3 & -1 \\ -3 & 0 & 0.5 \\ 1 & -0.5 & 0 \end{bmatrix}.$$

Note that this decomposition is unique.

A second-order tensor $\boldsymbol{S}$ is positive definite if $S_{ij}u_iv_j > 0$ for all vectors $\boldsymbol{u}, \boldsymbol{v} \in V_3$. Other conditions for positive definiteness are introduced later in this chapter.

As already mentioned, it is important to note that (3×1) and (3×3) matrices do not in general represent the components of first- or second-order tensors, respectively. The nine quantities S_{ik} are the components of a second-order tensor if, under orthogonal transformation of coordinates given by equations (1.8) and (1.10), they transform according to equations (1.30) and (1.31). If the transformation matrix $[\boldsymbol{a}]$ with elements a_{ik} and the square matrix $[\boldsymbol{S}]$ of tensor components S_{ij} are known, the matrix $[\boldsymbol{S}']$ with elements S'_{kl} can be obtained from the matrix form $[\boldsymbol{S}'] = [\boldsymbol{a}][\boldsymbol{S}][\boldsymbol{a}]^T$ of equation (1.31) by using Mathematica or a similar package. Matrix notation is sometimes useful in carrying out algebraic manipulations that involve components of vectors and second-order tensors. In Table 1.2 we list a number of examples of vector and tensor relations expressed in symbolic, suffix, and matrix notation.

Table 1.2. *Examples of symbolic, suffix, and matrix notation*

Symbolic notation	Suffix notation	Matrix notation
$\boldsymbol{u} \cdot \boldsymbol{v}$	$u_i v_i$	$[\boldsymbol{u}]^T[\boldsymbol{v}]$
$\boldsymbol{u} \otimes \boldsymbol{v}$	$u_i v_j$	$[\boldsymbol{u}][\boldsymbol{v}]^T$
$\boldsymbol{A}\boldsymbol{u}$	$A_{ij}u_j$	$[\mathbf{A}][\boldsymbol{u}]$
$\boldsymbol{A}^T\boldsymbol{u}$	$u_i A_{ij}$	$[\mathbf{A}]^T[\boldsymbol{u}]$
$\boldsymbol{u}^T\boldsymbol{A}\boldsymbol{v}$	$u_i A_{ij} v_j$	$[\boldsymbol{u}]^T[\mathbf{A}][\boldsymbol{v}]$
$\boldsymbol{A}\boldsymbol{B}$	$A_{ik}B_{kj}$	$[\mathbf{A}][\boldsymbol{B}]$
$\boldsymbol{A}\boldsymbol{B}^T$	$A_{ik}B_{jk}$	$[\mathbf{A}][\boldsymbol{B}]^T$

The inner or scalar product of the vectors $\boldsymbol{w}$ and $\boldsymbol{S}\boldsymbol{u}$ obeys the commutative rule and is denoted by

$$\boldsymbol{w} \cdot \boldsymbol{S}\boldsymbol{u} \text{ or } \boldsymbol{S}\boldsymbol{u} \cdot \boldsymbol{w}. \tag{1.33}$$

We must also note that

$$(\boldsymbol{S}\boldsymbol{u}) \cdot \boldsymbol{w} \neq \boldsymbol{S}(\boldsymbol{u} \cdot \boldsymbol{w}), \tag{1.34}$$

since the left-hand side is a scalar and the right-hand side is a tensor. Also we note that

$$\boldsymbol{w} \cdot \boldsymbol{S}\boldsymbol{u} = S_{ij}w_i u_j. \tag{1.35}$$

The product, not to be confused with the dyadic product, of two second-order tensors $\boldsymbol{S}$ and $\boldsymbol{T}$ is a second-order tensor with components

$$(\boldsymbol{S}\boldsymbol{T})_{ij} = S_{ik}T_{kj}. \tag{1.36}$$

This is essentially matrix multiplication. It is easily shown that the product of two second-order tensors is a second-order tensor. It should be noted that in general

$$\boldsymbol{S}\boldsymbol{T} \neq \boldsymbol{T}\boldsymbol{S},$$

that is, the product of two second-order tensors does not, in general, satisfy the commutative law.

The components of the unit tensor $\boldsymbol{I}$ are given by the Kronecker delta so that

$$\boldsymbol{I} = \delta_{ij}\mathbf{e}_i \otimes \mathbf{e}_j. \tag{1.37}$$

The unit tensor is an isotropic tensor, that is, it has the same components referred any coordinate system since

$$\delta'_{ij} = a_{ik}a_{jl}\delta_{kl} = a_{ik}a_{jk} = \delta_{ij},$$

where the substitution property of the Kronecker delta has been employed. It may also be shown that

$$\boldsymbol{I} = \mathbf{e}_i \otimes \mathbf{e}_i. \tag{1.38}$$

We note that in symbolic notation equation (1.6) becomes

$$\boldsymbol{u} = \boldsymbol{I}\boldsymbol{u}, \tag{1.39}$$

and

$$\boldsymbol{S} = \boldsymbol{I}\boldsymbol{S} = \boldsymbol{S}\boldsymbol{I}. \tag{1.40}$$

The inner product of two second-order tensors is a scalar defined by

$$\mathrm{tr}\left(\boldsymbol{S}^{\mathrm{T}}\boldsymbol{T}\right) = S_{ij}T_{ij}, \tag{1.41}$$

where tr denotes the trace and has the same significance as in linear algebra so that $\mathrm{tr}\boldsymbol{S} = S_{ii}$, that is, it is the sum of the diagonal elements of the component matrix. The magnitude of a second-order tensor is a scalar defined by,

$$|\boldsymbol{S}| = \{S_{ij}S_{ij}\}^{1/2} = \left\{\mathrm{tr}\left(\boldsymbol{S}^{\mathrm{T}}\boldsymbol{S}\right)\right\}^{1/2}. \tag{1.42}$$

It is shown in section 11 of this chapter that $\boldsymbol{S}^T\boldsymbol{S}$ is symmetric and positive or positive semidefinite. Consequently, $\mathrm{tr}(\mathbf{S}^T\mathbf{S}) \geqslant 0$.
It may be shown by using suffix notation that

$$(\boldsymbol{ST})^{\mathrm{T}} = \boldsymbol{T}^{\mathrm{T}}\boldsymbol{S}^{\mathrm{T}}. \tag{1.43}$$

A second-order tensor $\boldsymbol{C}$ has an inverse denoted by $\boldsymbol{C}^{-1}$, defined by

$$\boldsymbol{C}^{-1}\boldsymbol{C} = \boldsymbol{C}\boldsymbol{C}^{-1} = \boldsymbol{I}, \tag{1.44}$$

unless $\boldsymbol{C}$ is singular, that is, $\det[C] = 0$. If $\det[\boldsymbol{C}] \neq 0$, the tensor is said to be invertible. A further discussion of the inverse of a second-order tensor is deferred until section 10 of this chapter.

The inner product of $\mathbf{e}_i \otimes \mathbf{e}_j$ and $\mathbf{e}_k \otimes \mathbf{e}_t$ is

$$\mathrm{tr}\left[(\mathbf{e}_i \otimes \mathbf{e}_j)^{\mathrm{T}}(\mathbf{e}_k \otimes \mathbf{e}_t)\right] = \delta_{ik}\delta_{jt}, \tag{1.45}$$

and a simple application of the tensor transformation rule shows that the sum of second-order tensors is a second-order tensor. Consequently, the set of all second-order tensors is a nine-dimensional linear vector space with orthonormal basis $\mathbf{e}_i \otimes \mathbf{e}_j$, where the term vector is taken in its more general sense as belonging to a linear vector space.

It is often convenient to decompose a second-order tensor $\boldsymbol{T}$ into two parts, a deviatoric part and an isotropic part, as follows

$$T_{ij} = T^*_{ij} + \hat{T}\delta_{ij}, \tag{1.46}$$

where $\hat{T} = T_{kk}/3 = \mathrm{tr}T/3$. The deviatoric part with components T^*_{ij} has the property, $\mathrm{tr}\boldsymbol{T}^* = T^*_{ii} = 0$, and the isotropic part $\hat{T}\delta_{ij}$ has the same components in all coordinate systems. The decomposition equation (1.46) is unique and is important in linear elasticity and plasticity theory.

The concept of tensor multiplication can be extended to more than two vectors or to higher-order tensors. For example, the tensor product $\boldsymbol{T} \otimes \boldsymbol{S}$ of two second-order tensors $\boldsymbol{T}$ and $\boldsymbol{S}$ is a fourth-order tensor with components $T_{ij}S_{kl}$. A detailed discussion of higher-order tensors is given in section 9 of this chapter.

1.6 Tensor Algebra

The following results are easily verified by using the transformation rule.

1. Tensor addition. The sum or difference of two tensors of the same order is a tensor, also of the same order. For example, let A_{ij} and B_{ij} be the components, referred to the same coordinate system, of two second-order tensors. Then

$$C_{ij} = A_{ij} \pm B_{ij} \tag{1.47}$$

   are the components of a second-order tensor. It follows from equation (1.47) that addition of two tensors is commutative.

2. Tensor multiplication. We have already seen an example of this, namely, the tensor product of two vectors that results in a second-order tensor. A generalization of this is that the tensor product of a tensor of order m and a tensor of order n is a tensor of order $m + n$.

3. Contraction. We have already met an example of contraction. If we take the trace of $\boldsymbol{u} \otimes \boldsymbol{v}$, this is the same as contracting $u_i v_j$ and a scalar $u_i v_i$ is obtained so that

$$\boldsymbol{u} \cdot \boldsymbol{v} = \operatorname{tr}(\boldsymbol{u} \otimes \boldsymbol{v}). \tag{1.48}$$

Equation (1.48) is a result of a special case of contraction. In general, contraction involves two suffixes equating and summing over the resulting repeated suffix. A further special case is the contraction of a second-order tensor $\boldsymbol{S}$, with components S_{ij}, to obtain the scalar $\operatorname{tr}\boldsymbol{S} = S_{ii}$. Consider the third-order tensor with components A_{ijk}. It follows that A_{iik}, A_{ijj}, and A_{iji} represent the components of vectors. The concept can be extended to consider higher-order tensors. Contraction of two suffixes of an mth-order tensor (m $\geqslant$ 2) results in a tensor of order $m - 2$.

Test for tensor character. If u_i is a set of three quantities possessing the property that $u_i v_i$ is an invariant scalar for an arbitrary vector $\boldsymbol{v}$, then u_i

are the components of a vector. The reader should prove this. If C_{ij} is a set of nine quantities possessing the property that $C_{ij}u_j$ is a first-order tensor for an arbitrary vector $\boldsymbol{u}$, then C_{ij} are the components of a second-order tensor. Extension to consider higher-order quantities is straightforward.

1.7 Invariants of Second-Order Tensors

The theory in this section is closely related to the corresponding theory of square and column matrices in linear algebra.

The direction of the unit vector $\boldsymbol{n}$ is a principal direction of the symmetric tensor $\boldsymbol{A}$ if the vector $\boldsymbol{An}$ with components $A_{ij}n_j$ is parallel to $\boldsymbol{n}$, that is, if

$$A_{ij}n_j = \lambda n_i. \tag{1.49}$$

If $\lambda = 0$, the unit vector can only be obtained if $\boldsymbol{A}$ is singular, that is, if $\det[\boldsymbol{A}] = 0$. Equation (1.49) can be written as

$$\left(A_{ij} - \lambda\delta_{ij}\right)n_j = 0, \tag{1.50}$$

which represents a system of three linear homogeneous equations for the components n_1, n_2, and n_3 of $\boldsymbol{n}$. The trivial solution, $n_1 = n_2 = n_3 = 0$, of (1.50) is not compatible with the condition $n_i n_i = \boldsymbol{n}\cdot\boldsymbol{n} = 1$; consequently, the determinant of the coefficients in equation (1.50) must vanish if the system has a solution. That is, the requirement for a solution is

$$\det[\boldsymbol{A} - \lambda\boldsymbol{I}] = 0, \tag{1.51}$$

where $\det[\boldsymbol{A} - \lambda\boldsymbol{I}]$ denotes the determinant of the matrix of the coefficients of the tensor $\boldsymbol{A} - \lambda\boldsymbol{I}$. Equation (1.51) is known as the characteristic equation and is a cubic in λ, which may be written as

$$\lambda^3 - J_1\lambda^3 + J_2\lambda - J_3 = 0, \tag{1.52}$$

where

$$J_1 = A_{ii} = \operatorname{tr}\boldsymbol{A},\ J_2 = \frac{1}{2}\left[(A_{ii})^2 - A_{ik}A_{ki}\right] = \frac{1}{2}\left[(\operatorname{tr}\boldsymbol{A})^2 - \operatorname{tr}(\boldsymbol{A}^2)\right],\ J_3 = \det[A].$$

The coefficients J_1, J_2, and J_3, of the cubic equation (1.52), are scalar invariants and are known as the basic invariants of the tensor $\mathbf{A}$. It follows that the roots $(\lambda_1, \lambda_2, \lambda_3)$ of equation (1.52) are independent of the coordinate system and are the principal values of the symmetric tensor (or eigenvalues), and the corresponding directions are called the principal directions (or eigenvectors). The eigenvalues of the matrix of components are intrinsic to the tensor $\mathbf{A}$. If $\mathbf{A}$ is symmetric, the eigenvalues are real; this is sufficient but not necessary for real eigenvalues, and if the eigenvalues of $\mathbf{A}$, $(\lambda_1, \lambda_2, \lambda_3)$, are all positive, then A is a positive definite tensor. If $\mathbf{A}$ is symmetric, and if the eigenvalues are distinct, then the normalized eigenvectors $\left(\boldsymbol{n}^{(1)}, \boldsymbol{n}^{(2)}, \boldsymbol{n}^{(3)}\right)$ are mutually orthogonal. These two axioms are shown in the following manner. At least one of the roots of a cubic equation with real coefficients must be real; consequently, one of the principal values, say, λ_1 must be real. Let $\boldsymbol{n}^{(1)}$ be the unit vector associated with λ_1, which is obtained from the equations

$$\left(A_{ij} - \delta_{ij}\lambda_1\right)n_j^{(1)} = 0, \tag{1.53}$$

and the condition $\boldsymbol{n}^{(1)} \cdot \boldsymbol{n}^{(1)} = 1$. If the x_1' axis of a new coordinate system is in this direction, the tensor $\mathbf{A}$ has the primed components

$$A_{11}' = \lambda_1,\ A_{12}' = A_{13}' = 0,\ A_{22}',\ A_{13}',\ A_{23}'.$$

Consequently, the characteristic equation can be written as

$$\det\begin{bmatrix} \lambda_1 - \lambda & 0 & 0 \\ 0 & A_{22}' - \lambda & A_{23}' \\ 0 & A_{23}' & A_{33}' - \lambda \end{bmatrix} = 0$$

or

$$(\lambda_1 - \lambda)\left[\lambda^2 - (A_{22}' + A_{33}')\lambda + A_{22}'A_{33}' - A_{23}'^2\right] = 0. \tag{1.54}$$

This is known as the characteristic equation. The remaining principal values are the roots of

$$\left[\lambda^2 - (A'_{22} + A'_{33})\lambda + A'_{22}A'_{33} - A'^{2}_{23}\right] = 0, \tag{1.55}$$

and the discriminant of this quadratic equation is

$$(A'_{22} + A'_{33})^2 - 4\left(A'_{22}A'_{33} - A'^{2}_{23}\right) = (A'_{22} - A'_{33})^2 + 4\,A'^{2}_{23} \geqslant 0. \tag{1.56}$$

Consequently, the roots λ_2 and λ_3 are also real.

We will now show that if **A** is symmetric, and if λ_1, λ_2, and λ_3 are distinct, then the corresponding eigenvectors are mutually orthogonal. It follows from equation (1.53) that

$$A_{ij}n_j^{(1)} = \lambda_1 n_i^{(1)}. \tag{1.57}$$

Similarly

$$A_{ij}n_j^{(2)} = \lambda_2 n_i^{(2)}. \tag{1.58}$$

Scalar multiplication of equation (1.57) by $n_i^{(2)}$ and equation (1.58) by $n_j^{(1)}$ and subtracting gives

$$(\lambda_1 - \lambda_2)n_i^{(1)}n_i^{(2)} = 0,$$

which shows that $\boldsymbol{n}^{(1)}$ is orthogonal to $\boldsymbol{n}^{(2)}$ since $\lambda_1 \neq \lambda_2$.

If the tensor is referred to axes $0x''_i$ that coincide with the principal directions $\boldsymbol{n}^{(1)}$ the transformation matrix is given by

$$a'_{ij} = n_j^{(i)} = \boldsymbol{n}^{(i)} \cdot \mathbf{e}_j,$$

and $x_i'' = a_{ij}' x_j$. The components A_i'' referred to $0x_i''$ are then

$$A_i'' = n_k^{(i)} n_t^{(j)} A_{kt}, \tag{1.59}$$

and it can be deduced that

$$A_{ij}'' = \begin{bmatrix} \lambda_1 & 0 & 0 \\ 0 & \lambda_2 & 0 \\ 0 & 0 & \lambda_3 \end{bmatrix}.$$

This is known as the canonical form of the matrix of components. For many problems and proofs, it is desirable to refer components of second-order symmetric tensors to the principal axes. Equation (1.59) can be expressed in symbolic form

$$\text{diag}(\boldsymbol{A}) = \boldsymbol{N}\boldsymbol{A}\boldsymbol{N}^{\text{T}}, \tag{1.60}$$

where diag($\boldsymbol{A}$) is the canonical form of the matrix of components of $\boldsymbol{A}$ and $\boldsymbol{N}$ is a proper orthogonal second-order tensor with components

$$N_{ij} = n_j^{(i)}.$$

The rows of the matrix $\boldsymbol{N}$ are the eigenvectors of $\boldsymbol{A}$. Usually, in treatises on linear algebra, the diagonalization matrix is the transpose of $\boldsymbol{N}$, that is, the columns are the eignvectors of $\boldsymbol{A}$. The motivation for the above definition of $\boldsymbol{N}$ is that Mathematica gives a matrix whose rows are the eigenvectors.

If exactly two eigenvalues are equal, that is, $\lambda_1 \neq \lambda_2 = \lambda_3$, then the discriminant equation (1.56) is zero, that is, $A_{22}' = A_{33}'$, and $A_{23}' = 0$. Equation (1.54) then has roots $\lambda_2 = \lambda_3 = A_{22}' = A_{33}'$, and the matrix of components referred to the $0x_i'$ axes has a diagonal form regardless of the choice of the orthogonal pair of axes, $0x_2'$ and $0x_3'$ if the $0x_1'$ axis is in the $\boldsymbol{n}^{(1)}$ direction. It follows that any axis normal to $\boldsymbol{n}^{(1)}$ is a principal axis (eigenvector). As an exercise consider the case $\lambda_1 = \lambda_2 = \lambda_3$.

If the λ_i are distinct, the symmetric tensor $\boldsymbol{A}$ may be expressed, with the summation convention suspended, in the form,

$$\boldsymbol{A} = \sum_{i=1}^{3} \lambda_i \boldsymbol{n}^{(i)} \otimes \boldsymbol{n}^{(i)},$$

which is known as the spectral form. If two eigenvalues are equal, that is, if $\lambda_1 = \lambda_2 = \lambda_3$, then

$$\boldsymbol{A} = \lambda_1 \boldsymbol{n}^{(1)} \otimes \boldsymbol{n}^{(1)} + \lambda_2\left(\boldsymbol{I} - \boldsymbol{n}^{(1)} \otimes \boldsymbol{n}^{(1)}\right).$$

Further, if all three eigenvalues are equal, that is, if $\lambda_1 = \lambda_2 = \lambda_3 = \lambda$, then

$$\boldsymbol{A} = \lambda \boldsymbol{I},$$

and $\boldsymbol{A}$ is an isotropic tensor.

The eigenvalues and eigenvectors of a second-order tensor can be obtained using Mathematica and routine but tedious computations are avoided.

1.8 Cayley-Hamilton Theorem

The Cayley-Hamilton theorem has important applications in continuum mechanics. According to this theorem, a square matrix satisfies, in a matrix sense, its own characteristic equation. We will consider this theorem for a second-order symmetric tensor $\boldsymbol{C}$, although it is valid for any second-order tensor or square matrix. As shown in the previous section, the tensor $\boldsymbol{C}$ has three real eigenvalues λ_1, λ_2, and λ_3, which are the roots of the characteristic equation (1.54). Consequently, we have the matrix equation

$$[\lambda]^3 - J_1[\lambda]^2 + J_2[\lambda] - J[I] = [0],$$

where $[\boldsymbol{I}]$ is the unit (3×3) matrix and $[\lambda]$ is the diagonal matrix of the components of $\boldsymbol{C}$ referred to the principal axes. Since a tensor equation referred to a particular coordinate system is true referred to any other coordinate system, it may be deduced from equation (1.52) that

$$\boldsymbol{C}^3 - J_1\boldsymbol{C}^2 + J_2\boldsymbol{C} - J_3\boldsymbol{I} = 0, \tag{1.61}$$

which is the Cayley-Hamilton theorem for a second-order symmetric tensor. The Cayley-Hamilton theorem is also valid for any (3×3) matrix.

1.9 Higher-Order Tensors

The concept of tensor mutiplication can be extended to more than two vectors. For example, $(\boldsymbol{u} \otimes \boldsymbol{v} \otimes \boldsymbol{w})_{ijk} = u_i v_j w_k$, which transforms under coordinate transformation as

$$u'_r v'_s w'_t = a_{ri} a_{sj} a_{tk} u_i v_j w_k, \tag{1.62}$$

$$u_i v_j w_k = a_{ri} a_{sj} a_{tk} u' r v'_s w'_t. \tag{1.63}$$

A third-order tensor with components D_{ijk} may be expressible in the form

$$D_{ijk}\mathbf{e}_i \otimes \mathbf{e}_j \otimes \mathbf{e}_k,$$

which is invariant under coordinate transformation. The components transform according to

$$D'_{ijk} = a_{ir} a_{js} a_{kt} D_{rst}, \; D_{\mathrm{ijk}} = a_{\mathrm{ir}} a_{\mathrm{js}} a_{\mathrm{kt}} D'_{rst}.$$

The extension to higher-order tensors is obvious. The tensor product $\boldsymbol{S} \otimes \boldsymbol{T}$ of the two second-order tensors $\boldsymbol{S}$ and $\boldsymbol{T}$ is a fourth-order tensor, with components

$$(S \otimes T)_{ijrs} = S_{ij} T_{rs}. \tag{1.64}$$

We meet fourth-order tensors in the study of classical elasticity. It is sometimes convenient to consider a fourth-order tensor to be a linear transformation that assigns to each second-order tensor another second-order tensor. If B_{ijkt} are the components of a fourth-order tensor, then

$$S_{ij} = B_{ijkt}T_{kt},$$

where $\boldsymbol{T}$ is any second-order tensor and $\boldsymbol{S}$ is the result of $\boldsymbol{B}$ operating on $\boldsymbol{T}$ as indicated.

Symbolic notation is not particularly useful for tensors of higher-order than two and henceforth we use only suffix notation for these.

The alternating tensor, which is actually a pseudo tensor, is a useful third-order isotropic tensor if referred to right- (or left-) handed co-ordinate systems exclusively. The alternating symbol e_{ijk} is defined as: $\mathrm{e}_{ijk} = 1$ if (i, j, k) is an even permutation of (1, 2, 3), that is, if

$$\mathrm{e}_{123} = \mathrm{e}_{231} = \mathrm{e}_{312} = 1;$$

$e_{ijk} = -1$ if (i, j, k) is an odd permutation of (1, 2, 3), that is, if

$$\mathrm{e}_{321} = \mathrm{e}_{132} = \mathrm{e}_{213} = -1;$$

$e_{ijk} = 0$ if any two of i, j, k are equal (e.g., $\mathrm{e}_{112} = 0$, $\mathrm{e}_{333} = 0$).

The vector product of two base vectors is given by

$$\mathbf{e}_i \times \mathbf{e}_j = \mathrm{e}_{ijk}\mathbf{e}_k,$$

which follows from the definition of vector product in elementary vector algebra and the orthogonality of the unit base vectors. It then follows that the vector product of two vectors $\boldsymbol{u}$ and $\boldsymbol{v}$ can be expressed in the form

$$\boldsymbol{u} \times \boldsymbol{v} = u_i v_j \mathbf{e}_i \times \mathbf{e}_j = \mathrm{e}_{ijk} u_i v_j \mathbf{e}_k. \tag{1.65}$$

Since $\boldsymbol{u} \times \boldsymbol{v}$ is invariant under right-handed to right-handed (or left-handed to left-handed) coordinate transformations, it may be deduced from equation (1.65) that e_{ijk} may be regarded as the components of a third-order isotropic tensor for these transformations.

A useful identity is

$$\mathrm{e}_{ijk}\mathrm{e}_{ipq} = \delta_{jp}\delta_{kq} - \delta_{jq}\delta_{kp}. \tag{1.66}$$

In order to prove equation (1.66) we note that

$$(\boldsymbol{u} \times \boldsymbol{v}) \cdot (\boldsymbol{s} \times \boldsymbol{t}) = \mathrm{e}_{ijk}\mathrm{e}_{ipq}u_j v_k s_p t_q,$$

and

$$\begin{aligned} \boldsymbol{u} \times \boldsymbol{v} &= (u_1\mathbf{e}_1 + u_2\mathbf{e}_2 + u_3\mathbf{e}_3) \times (v_1\mathbf{e}_1 + v_2\mathbf{e}_2 + v_3\mathbf{e}_3) \\ &= \mathbf{e}_1(u_2v_3 - v_2\mathrm{u}_3) + \mathbf{e}_2(u_3v_1 - v_3u_1) + \mathbf{e}_3(u_1v_2 - v_1u_2). \end{aligned}$$

Similarly

$$\boldsymbol{s} \times \boldsymbol{t} = \mathbf{e}_1(s_2t_3 - t_2s_3) + \mathbf{e}_2(s_3t_1 - t_3s_1) + \mathbf{e}_3(s_1t_2 - t_1s_2).$$

It then follows that

$$\begin{aligned} (\boldsymbol{u} \times \boldsymbol{v}) \cdot (\boldsymbol{s} \times \boldsymbol{t}) &= (u_2v_3 - u_3v_2)(s_2t_3 - s_3t_2) + (u_3v_1 - u_1v_3)(s_3t_1 - s_1t_3) \\ &\quad + (u_1v_2 - u_2v_1)(s_1t_2 - s_2t_1) \\ &= \left(\delta_{jp}\delta_{kq} - \delta_{jq}\delta_{kp}\right)u_jv_ks_pt_q. \end{aligned}$$

Comparison of the above two expressions for $(\boldsymbol{u} \times \boldsymbol{v}) \cdot (\boldsymbol{s} \times \boldsymbol{t})$ gives equation (1.66). The identity equation (1.66) can be used to obtain the relation

$$\boldsymbol{a} \times (\boldsymbol{b} \times \boldsymbol{c}) = \boldsymbol{b}(\boldsymbol{a} \cdot \boldsymbol{c}) - \boldsymbol{c}(\boldsymbol{a} \cdot \boldsymbol{b}).$$

This is obtained from equations (1.65) and (1.66) as follows,

$$\begin{aligned} \boldsymbol{a} \times (\boldsymbol{b} \times \boldsymbol{c}) &= \mathrm{e}_{ijk}a_j(\mathrm{e}_{lmk}b_lc_m)\mathbf{e}_i \\ &= \left(\delta_{il}\delta_{jm} - \delta_{im}\delta_{lj}\right)a_jb_lc_m\mathbf{e}_i \\ &= (b_ia_mc_m - c_ia_lb_l)\mathbf{e}_i \\ &= \boldsymbol{b}(\boldsymbol{a} \cdot \boldsymbol{c}) - \boldsymbol{c}(\boldsymbol{a} \cdot \boldsymbol{b}). \end{aligned}$$

It should be noted that, in general, $\boldsymbol{a} \times (\boldsymbol{b} \times \boldsymbol{c}) \neq (\boldsymbol{a} \times \boldsymbol{b}) \times \boldsymbol{c}$.

Identity equation (1.66) involves fourth-order tensors, and it may be deduced that this is an isotropic tensor. This tensor is a particular case of the most general fourth-order isotropic tensor that has components

$$\alpha\delta_{ij}\delta_{kl} + \beta\delta_{ik}\delta_{jl} + \gamma\delta_{il}\delta_{jk}, \tag{1.67}$$

where α, β, and γ are scalars.

1.10 Determinants

The Laplace expansion of a determinant can be expressed in the form

$$\det\begin{bmatrix} A_1 & A_2 & A_3 \\ B_1 & B_2 & B_3 \\ C_1 & C_2 & C_3 \end{bmatrix} = \mathrm{e}_{ijk}A_iB_jC_k, \tag{1.68}$$

where $\mathbf{e}_{ijk}$ are the components of the alternating tensor as given in section 9.

It follows from the Laplace expansion that $\det[\boldsymbol{I}] = 1$ and from equations (1.65) and (1.68) that the vector product of two vectors $\boldsymbol{u}$ and $\boldsymbol{v}$ can be expressed as a determinant,

$$\boldsymbol{u} \times \boldsymbol{v} = \det\begin{bmatrix} \mathbf{e_1} & \mathbf{e_2} & \mathbf{e_3} \\ u_1 & u_2 & u_3 \\ v_1 & v_2 & v_3 \end{bmatrix},$$

The triple scalar product $\boldsymbol{u} \cdot \boldsymbol{v} \times \boldsymbol{w}$ is given by,

$$\boldsymbol{u} \cdot \boldsymbol{v} \times \boldsymbol{w} = \det\begin{bmatrix} u_1 & u_2 & u_3 \\ v_1 & v_2 & v_3 \\ w_1 & w_2 & w_3 \end{bmatrix} = \mathrm{e}_{ijk}u_iv_jw_k. \tag{1.69}$$

A physical significance of the triple scalar product is that equation (1.69) represents the enclosed volume of the parallelpiped with contiguous sides of the three noncoplanar vectors, $\boldsymbol{u}$, $\boldsymbol{v}$, and $\boldsymbol{w}$. We will use this interpretation in later chapters.

It follows from equation (1.69) that

$$\det\begin{bmatrix} C_{11} & C_{12} & C_{13} \\ C_{21} & C_{22} & C_{23} \\ C_{31} & C_{32} & C_{33} \end{bmatrix} = \mathrm{e}_{ijk} C_{1i} C_{2j} C_{3k} \text{ expanded by rows} \\ = \mathrm{e}_{ijk} C_{i1} C_{j2} C_{k3} \text{ expanded by columns.} \tag{1.70}$$

Further important relations are,

$$\mathrm{e}_{ijk} C_{ir} C_{js} C_{kt} = \mathrm{e}_{rst} \det[C], \tag{1.71}$$

$$\mathrm{e}_{ijk} C_{ri} C_{sj} C_{tk} = \mathrm{e}_{rst} \det[C], \tag{1.72}$$

$$\frac{1}{6} \mathrm{e}_{ijk} \mathrm{e}_{rst} C_{ir} C_{js} C_{kt} = \mathrm{e}_{rst} \det[C]. \tag{1.73}$$

In equations (1.70) to (1.73) we assume that C_{ij} are the components of the second-order tensor $\boldsymbol{C}$, although the results are valid for any (3×3) matrix. The determinant of a second-order tensor is a scalar and is known as the third invariant of the tensor, as already mentioned in section 7. If $\det[\boldsymbol{C}] = 0$, the tensor is said to be singular. The coefficient of C_{ij} in the expansion of the determinant is called the cofactor of C_{ij} and is denoted by C^c_{ij}, so that expansion by rows gives

$$\det[\boldsymbol{C}] = C_{1j} C^c_{1j} = C_{2j} C^c_{2j} = C_{3j} C^c_{3j},$$

which in turn gives

$$C_{ij} C^c_{kj} = \delta_{ik} \det[\boldsymbol{C}]. \tag{1.74}$$

It may be deduced from equation (1.74) that C^c_{kj} are the components of a second-order tensor; consequently, we can express equation (1.74) in symbolic notation as,

$$\boldsymbol{C}\boldsymbol{C}^{cT} = \boldsymbol{I} \det[\boldsymbol{C}], \tag{1.75}$$

where the matrix of components of $\boldsymbol{C}^{cT}$ is the transpose of the matrix of the cofactors of $\boldsymbol{C}$. It then follows from equation (1.74) that

$$\mathrm{tr}\boldsymbol{C}\boldsymbol{C}^{cT} = C_{ij}C^{c}_{ij} = 3\det[\boldsymbol{C}].$$

EXAMPLE PROBLEM 1.3 Show that

$$C^{c}_{mi} = \frac{1}{2}\mathrm{e}_{ijk}\mathrm{e}_{imst}C_{sj}C_{tk}.$$

SOLUTION. Multiply each side of equation (1.71) by $\mathbf{e}_{mst}$ to obtain

$$\mathrm{e}_{mst}\mathrm{e}_{ijk}C_{ri}C_{sj}C_{tk} = \mathrm{e}_{mst}\mathrm{e}_{rst}\det[C].$$

It follows from equation (1.66) that $\mathrm{e}_{mst}\mathrm{e}_{rst} = 2\delta_{mr}$; consequently,

$$\mathrm{e}_{mst}\mathrm{e}_{ijk}C_{ri}C_{sj}C_{tk} = 2\delta_{rm}\det[C].$$

Compare with $C_{ri}C^{c}_{mi} = \delta_{rm}\det[\boldsymbol{C}]$ to obtain $C^{c}_{mi} = \frac{1}{2}\mathrm{e}_{ijk}\mathrm{e}_{mst}C_{sj}C_{tk}$.

The inverse of $\boldsymbol{C}$ is a second-order tensor denoted by $\boldsymbol{C}^{-1}$ and

$$\boldsymbol{C}\boldsymbol{C}^{-1} = \boldsymbol{I} \; or \; C_{ij}C^{-1}_{\mathrm{jk}} = \delta_{ik}. \tag{1.76}$$

If the tensor $\boldsymbol{C}$ is singular, the inverse does not exist. The term inverse is meaningful for scalars and second-order tensors but not for tensors of any other order. The relation $(\boldsymbol{CB})(\boldsymbol{CB})^{-1} = \boldsymbol{I}$ follows from equation (1.76), and premultipying each side first by $\boldsymbol{C}^{-1}$ and then by $\boldsymbol{B}^{-1}$ gives the the result

$$(\boldsymbol{CB})^{-1} = \boldsymbol{B}^{-1}\boldsymbol{C}^{-1}.$$

It follows from equation (1.72) that

$$C_{\mathrm{ij}}^{-1} = \frac{C_{ji}^{c}}{\det[\boldsymbol{C}]}. \tag{1.77}$$

Since the value of a determinant is unchanged if the rows and columns are interchanged.

$$\det[\boldsymbol{C}] = \det\left[\boldsymbol{C}^{T}\right]. \tag{1.78}$$

EXAMPLE PROBLEM 1.4. Prove that

$$\boldsymbol{Au} \cdot \boldsymbol{Av} \times \boldsymbol{Aw} = \det[\boldsymbol{A}](\boldsymbol{u} \cdot \boldsymbol{v} \times \boldsymbol{w}).$$

SOLUTION.

$$\begin{aligned}\boldsymbol{Au} \cdot \boldsymbol{Av} \times \boldsymbol{Aw} &= \mathrm{e}_{ijk} A_{ir} u_r A_{js} v_s A_{kt} w_t \\ &= \mathrm{e}_{rst} \det[A] u_r v_s w_t = (\boldsymbol{u} \cdot \boldsymbol{v} \times \boldsymbol{w}) \det[\boldsymbol{A}].\end{aligned}$$

It may be deduced from equation (1.70) that

$$\det[\alpha \boldsymbol{C}] = \alpha^{3} \det[\mathbf{C}],$$

and from the result of example problem (1.4) that

$$\det[\boldsymbol{CB}] = \det[\boldsymbol{C}]\det[\boldsymbol{B}], \tag{1.79}$$

since replacing $\boldsymbol{A}$ with $\boldsymbol{BC}$ gives

$$\begin{aligned}\boldsymbol{CBu} \cdot \boldsymbol{CBv} \times \boldsymbol{CBw} &= \det[\boldsymbol{CB}](\boldsymbol{u} \cdot \boldsymbol{v} \times \boldsymbol{w}) \\ &= \det[\boldsymbol{C}](\boldsymbol{Bu} \cdot \boldsymbol{Bv} \times \boldsymbol{Bw}) \\ &= \det[\boldsymbol{C}]\det[\boldsymbol{B}](\boldsymbol{u} \cdot \boldsymbol{v} \times \boldsymbol{w}),\end{aligned}$$

where $\boldsymbol{C}$ and $\boldsymbol{B}$ are nonsingular second-order tensors. It then follows from equation (1.79) by replacing $\boldsymbol{B}$ by $\boldsymbol{C}^{-1}$ that

$$\det[\boldsymbol{C}] = \left[\det\left[\boldsymbol{C}^{-1}\right]\right]^{-1}, \tag{1.80}$$

1.11 Polar Decomposition Theorem

We have already seen that an arbitrary second-order tensor can be decomposed, uniquely, into the sum of symmetric and antisymmetric parts. A further decomposition of an arbitrary nonsingular second-order tensor is given by the polar decomposition theorem. Before introducing this theorem we define the terms orthogonal second-order tensor and positive definite second-order tensor.

An orthogonal second-order tensor $\boldsymbol{Q}$ satisfies the condition,

$$\boldsymbol{Q}\boldsymbol{Q}^T = \boldsymbol{Q}^T\boldsymbol{Q} = \boldsymbol{I}, \tag{1.81}$$

or, in suffix notation,

$$\mathrm{Q}_{ji}Q_{ki} = Q_{ij}Q_{ik} = \delta_{\mathrm{jk}}.$$

It follows from equations (1.73) and (1.81) that

$$\boldsymbol{Q}^T = \boldsymbol{Q}^{-1},$$

$$\det[\boldsymbol{Q}] = \pm 1,$$

and $\boldsymbol{Q}$ is described as a proper (improper) orthogonal tensor and when the plus (minus) sign is valid. We are concerned with proper orthogonal tensors and henceforth use the term orthogonal tensor to denote a second-order tensor, which satisfies condition (1.81) and $\det[\boldsymbol{Q}] = 1$. A proper orthogonal tensor is similar, in a sense, to the transformation matrix for a proper orthogonal transformation. This is evident by comparing equations (1.19) and (1.20) with equation (1.81). The scalar product of two arbitrary vectors

$\boldsymbol{u}$, $\boldsymbol{v} \in V_3$ is unchanged if the vectors are operated on by an orthogonal tensor. That is,

$$\boldsymbol{u} \cdot \boldsymbol{v} = \boldsymbol{Qu} \cdot \boldsymbol{Qv}, \tag{1.82}$$

or, in suffix notation,

$$u_i v_i = Q_{ij} Q_{ik} u_j v_k.$$

This is sometimes given as the definition of an orthogonal tensor. Equation (1.82) follows from equation (1.77).

When an orthogonal tensor operates on a vector $\boldsymbol{u} \in V_3$, a vector of equal magnitude results, that is,

$$|\boldsymbol{u}| = |\boldsymbol{Qu}|. \tag{1.83}$$

Equation (1.83) is obtained from equation (1.82) by putting $\boldsymbol{u} = \boldsymbol{v}$. The unit tensor $\boldsymbol{I}$ is a special case of an orthogonal tensor.

A second-order tensor $\boldsymbol{S}$ is said to positive semidefinite if, for an arbitrary vector, $\boldsymbol{u}$,

$$\boldsymbol{u} \cdot \boldsymbol{Su} = \mathrm{S}_{ij} u_i u_j \geqslant 0, \tag{1.84}$$

and positive definite if the equality sign holds if and only if $\boldsymbol{u}$ is the null vector. The antisymmetric part of $\mathbf{S}$ does not contribute to $\boldsymbol{u} \cdot \boldsymbol{Su}$ since, if $\boldsymbol{W}$ is an antisymmetric second-order tensor,

$$\boldsymbol{u} \cdot \boldsymbol{Wu} = \mathrm{W}_{ij} u_i u_j = 0.$$

When we consider second-order positive definite tensors we usually assume the tensor is symmetric, and henceforth, the term positive definite tensor is used for a positive definite symmetric second-order tensor.

The second-order tensor $\boldsymbol{S} = \boldsymbol{A}^T\boldsymbol{A}$, where $\boldsymbol{A}$ is an arbitrary second-order tensor, is symmetric since $(\boldsymbol{A}^T\boldsymbol{A})^T = \boldsymbol{A}^T\boldsymbol{A}$. It then follows that $\boldsymbol{S}$ is either positive definite or semidefinite since

$$\boldsymbol{u}\cdot\boldsymbol{S}\boldsymbol{u} = \boldsymbol{u}\cdot\left(\boldsymbol{A}^T\boldsymbol{A}\right)\boldsymbol{u} = (\boldsymbol{A}\boldsymbol{u})\cdot(\boldsymbol{A}\boldsymbol{u}) \geqslant 0. \tag{1.85}$$

Inequality (1.85) is evident if it is expressed in suffix notation since

$$\boldsymbol{u}\cdot\left(\boldsymbol{A}^T\boldsymbol{A}\right)\boldsymbol{u} = u_jA_{ij}A_{ik}u_k = (A_{ij}u_j)(A_{ik}u_k).$$

If the tensor $\boldsymbol{A}$ is singular, that is, $\det[\boldsymbol{A}] = 0$, it may be deduced from equation (1.85) that $\boldsymbol{S}$ is positive semidefinite since $\boldsymbol{A}\boldsymbol{u} = 0$ has a nontrivial solution for $\boldsymbol{u}$. If $\boldsymbol{S}$ can be expressed in the form $\boldsymbol{S} = \boldsymbol{A}\boldsymbol{A}^T$ or $\boldsymbol{S} = \boldsymbol{A}^T\boldsymbol{A}$, it follows from equation (1.85) that a necessary and sufficient condition for $\boldsymbol{S}$ to be positive definite is $\det[\boldsymbol{S}] > 0$. If S is an arbitrary second-order symmetric tensor, the condition $\det[\boldsymbol{S}] > 0$ is necessary but not sufficient condition for positive definiteness. The necessary and sufficient condition for a symmetric second-order tensor to be positive definite is that the eigenvalues (or principal values) are all positive. This can be shown by expressing equation (1.84) in terms of the components with respect to the principal axes of $\boldsymbol{S}$, that is, axes, that are parallel to the eigenvectors (principal directions) of $\boldsymbol{S}$,

$$\boldsymbol{u}\cdot\boldsymbol{S}\boldsymbol{u} = \lambda_1\hat{u}_1^2 + \lambda_2\hat{u}_2^2 + \lambda_3\hat{u}_3^2$$

where λ_1, λ_2, λ_3 are the eigenvalues of $\boldsymbol{S}$ and $\hat{u}_1$, $\hat{u}_2$, and $\hat{u}_3$ are the components of $\boldsymbol{u}$ referred to the principal axes of $\boldsymbol{S}$. When all the eigenvalues of $\boldsymbol{S}$ are negative, it follows from equation (1.84) that $\boldsymbol{u}\cdot\boldsymbol{S}\boldsymbol{u} \leqslant 0$ with the equality sign holding only when $\boldsymbol{u}$ is the null vector. The tensor $\boldsymbol{S}$ is then said to be negative definite.

The square root $\boldsymbol{S}^{1/2}$ of a positive definite tensor $\boldsymbol{S}$ is defined as the symmetric second-order tensor whose spectral form is

$$\boldsymbol{S}^{1/2} = \sum_{i=1}^{3} \lambda_i^{1/2}\boldsymbol{n}^{(1)} \otimes \boldsymbol{n}^{(1)} \tag{1.86}$$

if the eigenvalues are distinct,

$$\begin{aligned} S^{1/2} &= \lambda_1^{1/2} n^{(1)} \otimes n^{(1)} + \lambda_2^{1/2}\left(I - n^{(1)} \otimes n^{(1)}\right) \\ if\, \lambda_2 &= \lambda_3 \text{ and} \\ S^{1/2} &= \lambda^{1/2} I \end{aligned}$$

if the eigenvalues are equal, that is, $\lambda_1 = \lambda_2 = \lambda_3 = \lambda$.

The summation convention is suspended and λ_1 and $n^{(i)}$, $i \in \{1, 2, 3\}$, are the eigenvalues and eigenvectors, respectively, of S. We adopt the convention that positive roots of the λ_1 are taken in equation (1.86) so that $S^{1/2}$ is positive definite and unique.

If the matrix of components of the positive definite tensor S is known, with respect to an arbitrary system of Cartesian axes, the matrix of components of $S^{1/2}$, with respect to the same system of axes, can be obtained as follows. First the eigenvalues $(\lambda_{(1)}, \lambda_{(2)}, \lambda_{(3)})$ of S are determined, then the square roots $\left(\lambda_1^{1/2}, \lambda_2^{1/2}, \lambda_3^{1/2}\right)$. It then follows from equation (1.60) that the matrix of components of $S^{1/2}$ is given by

$$[N]^T \begin{bmatrix} \lambda_1^{1/2} & 0 & 0 \\ 0 & \lambda_2^{1/2} & 0 \\ 0 & 0 & \lambda_3^{1/2} \end{bmatrix} [N],$$

where the first, second, and third rows of N are the eigenvectors of S corresponding to $\left(\lambda_1^{1/2}, \lambda_2^{1/2}, \lambda_3^{1/2}\right)$, respectively. It should be noted that the eigenvectors of $S^{1/2}$ are the same as those of S. The square root of a matrix of components of a second-order positive definite tensor can be determined directly by using Mathematica.

We are now ready to introduce the polar decomposition theorem, which states that an arbitrary nonsingular second-order tensor A with

$$\det[A] > 0$$

can be expressed uniquely in the forms,

$$A = QU, \tag{1.87}$$

$$A = VQ, \tag{1.88}$$

where $\boldsymbol{Q}$ is an orthogonal tensor and $\boldsymbol{U}$ and $\boldsymbol{V}$ are positive definite second-order tensors. If $\det[\boldsymbol{A}] < 0$, then $\boldsymbol{U}$ and $\boldsymbol{V}$ are negative definite tensors; however, we do not consider this case since it is not important in the applications that follow in later chapters. Equations (1.87) and (1.88) are known as the right and left polar decompositions, respectively. In order to prove the theorem, we note that $\boldsymbol{A}^{\mathrm{T}}\boldsymbol{A}$ and $\boldsymbol{A}\boldsymbol{A}^{\mathrm{T}}$ are positive definite tensors, so that $(\boldsymbol{A}^{\mathrm{T}}\boldsymbol{A})^{1/2}$ and $(\boldsymbol{A}\boldsymbol{A}^{\mathrm{T}})^{1/2}$ are also positive definite. The tensors $\boldsymbol{Q} = \boldsymbol{A}(\boldsymbol{A}^{\mathrm{T}}\boldsymbol{A})^{-1/2}$ and $\boldsymbol{P} = (\boldsymbol{A}\boldsymbol{A}^{\mathrm{T}})^{-1/2}\boldsymbol{A}$ are orthogonal since

$$\boldsymbol{A}\left(\boldsymbol{A}^{\mathrm{T}}\boldsymbol{A}\right)^{-1/2}\left[\boldsymbol{A}\left(\boldsymbol{A}^{\mathrm{T}}\boldsymbol{A}\right)^{-1/2}\right]^{\mathrm{T}} = \boldsymbol{A}\left(\boldsymbol{A}^{\mathrm{T}}\boldsymbol{A}\right)^{-1}\boldsymbol{A}^{\mathrm{T}}$$
$$= \boldsymbol{A}\boldsymbol{A}^{-1}\boldsymbol{A}^{\mathrm{T}^{-1}}\boldsymbol{A}^{\mathrm{T}} = \boldsymbol{I},$$

and $\det[\boldsymbol{P}] = \det[\boldsymbol{Q}] = 1$, so that $\boldsymbol{A} = \boldsymbol{Q}\boldsymbol{U} = \boldsymbol{V}\boldsymbol{P}$, where $\boldsymbol{U} = (\boldsymbol{A}^{\mathrm{T}}\boldsymbol{A})^{1/2}$ and $\boldsymbol{V} = (\boldsymbol{A}\boldsymbol{A}^{\mathrm{T}})^{1/2}$.

The square root of a positive definite tensor is unique, when the positive roots are taken in equation (1.83), so that the decomposition $\boldsymbol{A} = \boldsymbol{Q}\boldsymbol{U}$ is unique. We now have to show that $\boldsymbol{Q} = \boldsymbol{P}$ to complete the proof of the theorem. This follows since

$$\boldsymbol{A} = \boldsymbol{V}\boldsymbol{P} = \boldsymbol{P}\left(\boldsymbol{P}^{\mathrm{T}}\boldsymbol{V}\boldsymbol{P}\right),$$

and it can be shown that $\boldsymbol{P}^{\mathrm{T}}\boldsymbol{V}\boldsymbol{P} = \boldsymbol{U}$. Consequently, $\boldsymbol{Q} = \boldsymbol{P}$ and $\boldsymbol{U} = \boldsymbol{Q}^{\mathrm{T}}\boldsymbol{V}\boldsymbol{Q}$.

The positive definite tensors $\boldsymbol{U}$ and $\boldsymbol{V}$ have the same eigenvalues, and the eigenvectors $\boldsymbol{l}^{(i)}$ of $\boldsymbol{U}$ are related to the eigenvectors $\boldsymbol{m}^{(i)}$ of $\boldsymbol{V}$ by

$$\boldsymbol{l}^{(i)} = \boldsymbol{Q}^{\mathrm{T}}\boldsymbol{m}^{(i)}. \qquad (1.89)$$

EXAMPLE PROBLEM 1.5. Determine the polar decomposition equation (1.87) of the matrix of components, with respect to a particular coordinate system,

$$[\boldsymbol{A}] = \begin{bmatrix} 1.1 & 0.5 & 0.3 \\ 0.4 & 1.2 & 0.5 \\ -0.2 & 0.3 & 0.9 \end{bmatrix},$$

of a nonsingular second-order tensor $\boldsymbol{A}$.

SOLUTION. The matrix of components of

$$\boldsymbol{C} = \boldsymbol{A}^{\mathrm{T}}\boldsymbol{A}$$

is given by

$$[\boldsymbol{C}] = \begin{bmatrix} 1.1 & 0.4 & -0.2 \\ 0.5 & 1.2 & 0.3 \\ 0.3 & 0.5 & 0.9 \end{bmatrix} \begin{bmatrix} 1.1 & 0.5 & 0.3 \\ 0.4 & 1.2 & 0.5 \\ -0.2 & 0.3 & 0.9 \end{bmatrix}$$
$$= \begin{bmatrix} 1.41 & 0.97 & 0.35 \\ 0.97 & 1.78 & 1.02 \\ 0.35 & 1.02 & 1.15 \end{bmatrix}$$

and the matrix of components of

$$\boldsymbol{C}^{1/2} = \boldsymbol{U}$$

by

$$[\boldsymbol{U}] = \begin{bmatrix} 1.11294 & 0.406455 & 0.0785375 \\ 0.406455 & 1.18478 & 0.459452 \\ 0.0785375 & 0.459452 & 0.965783 \end{bmatrix},$$

which can be obtained by using Mathematica or the procedure already described.

The matrix of components of the inverse of $\boldsymbol{U}$

$$\left[\boldsymbol{U}^{-1}\right] = \begin{bmatrix} 1.036 & -0.3956 & 0.104 \\ -0.3956 & 1.186 & -0.532 \\ 0.104 & -0.532 & 1.280 \end{bmatrix}$$

can be obtained using Mathematica or a standard procedure given in most texts on linear algebra.

The matrix of components of the tensor

$$\boldsymbol{Q} = \boldsymbol{A}\boldsymbol{U}^{-1}$$

is

$$[\boldsymbol{Q}] = \begin{bmatrix} 1.1 & 0.5 & 0.3 \\ 0.4 & 1.2 & 0.5 \\ -0.2 & 0.3 & 0.9 \end{bmatrix} \begin{bmatrix} 1.036 & -0.3956 & 0.104 \\ -0.3956 & 1.186 & -0.532 \\ 0.104 & -0.532 & 1.280 \end{bmatrix}$$

$$= \begin{bmatrix} 0.9726 & 0.018 & 0.2324 \\ -0.0085 & 1.0 & 0.04314 \\ -0.2322 & -0.4393 & 0.9717 \end{bmatrix}.$$

It is easily verified that $[\boldsymbol{A}] = [\boldsymbol{Q}][\boldsymbol{U}]$.

1.12 Tensor Fields

If the value of a tensor is assigned, at every point $\boldsymbol{x}$ of a domain D of a Euclidean point space $\boldsymbol{E}_3$, we have a tensor field. The tensor, which may be of any order, may then be expressed as a function of $\boldsymbol{x} \in D$, and perhaps time. We are mainly concerned with scalar (zeroth-order tensor), vector (first-order tensor), and second-order tensor fields''. These fields may have discontinuities, but for the present we will assume that the tensor fields are continuous with continuous first derivatives. The treatment of derivatives of tensor components, with respect to spatial coordinates, is greatly simplified when rectangular Cartesian coordinates are adopted. In this section both suffix and symbolic notation are used to develop the theory.

First we consider a scalar field $\phi(\boldsymbol{x})$. Then $\partial\phi/\partial x_i$, sometimes written as $\phi_{,i}$, are the Cartesian components of a vector since,

$$\frac{\partial \phi}{\partial x'_j} = \frac{\partial \phi}{\partial x_i}\frac{\partial \phi}{\partial x'_j},$$

which, with the use of equation (1.11) becomes,

$$\frac{\partial \phi}{\partial x'_j} = a_{ji} \frac{\partial \phi}{\partial x_i}. \tag{1.90}$$

Equation (1.90) is the transformation rule for the vector $(\partial\phi/\partial x_i)\mathbf{e}_i \in \boldsymbol{V}_3$, which is known as the gradient of ϕ and is often denoted by $\nabla\phi$ or *grad*ϕ. At any point $\nabla\phi$ is normal to the surface $\phi(x_1, x_2, x_3) =$ constant passing through the point and is in the direction of ϕ increasing. The scalar $\nabla\phi \cdot \boldsymbol{n}$, where $\boldsymbol{n}$ is a unit vector, is the rate of change of ϕ with respect to the distance in the direction $\boldsymbol{n}$.

We use the symbol ∇ preceding an operand to denote either the differential operator

$$\nabla \equiv \mathrm{e}_i \frac{\partial()}{\partial x_i} \tag{1.91}$$

or

$$\nabla \equiv \frac{\partial()}{\partial x_i} \mathrm{e}_i, \tag{1.92}$$

and we will carefully note which definition is intended. It is immaterial which definition of ∇ is used for the gradient of a scalar or the divergence of a vector. When ∇ operates scalarly on a vector, it gives the divergence of the vector,

$$\nabla \cdot \boldsymbol{v} = \frac{\partial v_i}{\partial x_i}.$$

When ∇ operates scalarly, with either definition, on itself, the result is a scalar differential operator

$$(\nabla \cdot \nabla) \equiv \nabla^2 = \frac{\partial^2}{\partial x_i \partial x_i}.$$

According to the definition of curl $\boldsymbol{v}$, where $\boldsymbol{v}$ is a vector field, in elementary vector analysis,

$$\text{curl } \boldsymbol{v} = e_{ijk}\frac{\partial v_k}{\partial x_j}\mathbf{e}_i, \tag{1.93}$$

so that

$$\text{curl } \boldsymbol{v} \equiv \nabla \times \boldsymbol{v},$$

with ∇ given by definition (1.91). However, with definition (1.92), curl $\boldsymbol{v} \equiv - \nabla \times \boldsymbol{v}$.

Henceforth, the form

$$\nabla \times \boldsymbol{v}$$

is not used and curl $\boldsymbol{v}$ is understood to be given by equation (1.93).

The nine quantities $\partial v_i/\partial x_j$ where v_i are the components of a vector are the components of a second-order tensor field since it is easily shown that they transform according to the tensor transformation rule,

$$\frac{\partial v_t'}{\partial x_m'} = a_{tn}a_{mj}\frac{\partial v_n}{\partial x_j}.$$

The second-order tensor $\nabla \otimes \boldsymbol{v}$ with components $\partial v_i/\partial x_j$ is known as the gradient of vector $\boldsymbol{v}$, and the definition (1.92) is implied, so that

$$\nabla \otimes \boldsymbol{v} = \frac{\partial v_k}{\partial x_i}\mathbf{e}_k \otimes \mathbf{e}_i. \tag{1.94}$$

The matrix of components of $\nabla \otimes \boldsymbol{v}$ is then given by

$$\left[\frac{\partial v_i}{\partial x_j}\right] = \begin{bmatrix} \frac{\partial v_1}{\partial x_1} & \frac{\partial v_1}{\partial x_2} & \frac{\partial v_1}{\partial x_3} \\ \frac{\partial v_2}{\partial x_1} & \frac{\partial v_2}{\partial x_2} & \frac{\partial v_2}{\partial x_3} \\ \frac{\partial v_3}{\partial x_1} & \frac{\partial v_3}{\partial x_2} & \frac{\partial v_3}{\partial x_3} \end{bmatrix}. \tag{1.95}$$

It follows from equation (1.95) that $\nabla \cdot \boldsymbol{v} = tr(\nabla \otimes \boldsymbol{v})$. The notation $v_{k,i}$ is often used for $\partial v_k / \partial x_i$, and this is compatible with equations (1.94) and (1.95). When definition (1.92) is used we obtain the transpose of the right-hand side of equation (1.95). With definition (1.92),

$$d\boldsymbol{v} = (\nabla \otimes \boldsymbol{v})d\boldsymbol{x},$$

which is analogous to the result

$$d\mathbf{f}(\mathbf{x}) = \frac{d\mathbf{f}}{d\boldsymbol{x}}d\mathbf{x}$$

in elementary calculus, whereas with definition (1.91), we obtain

$$d\boldsymbol{v} = d\boldsymbol{x} \cdot (\nabla \otimes \boldsymbol{v}).$$

In spite of the difficulty with the curl it is usually more convenient to adopt definition (1.92). The gradient of the vector field $v(\boldsymbol{x})$ may also be defined as the linear transformation that assigns to a vector $\boldsymbol{a}$ the vector given by the following rule:

$$(\nabla \otimes \boldsymbol{v})\boldsymbol{a} = \lim_{s \to 0} \frac{\boldsymbol{v}(\boldsymbol{x} + s\boldsymbol{a}) - \boldsymbol{v}(\boldsymbol{x})}{s},$$

which is compatible with definition (1.92) and is analogous to a definition of the gradient of a scalar field $\phi(\boldsymbol{x})$, namely, that the gradient of $\phi(\boldsymbol{x})$ assigns to a vector $\boldsymbol{a}$ the scalar given by the following rule

$$(\nabla \phi) \cdot \boldsymbol{a} = \lim_{s \to 0} \frac{\phi(\boldsymbol{x} + s\boldsymbol{a}) - \phi(\mathbf{x})}{s}.$$

The divergence of a second-order tensor $\boldsymbol{T}$ is a vector given by

$$\text{div}\boldsymbol{T} = \frac{\partial T_{ij}}{\partial x_i}\mathbf{e}_j \tag{1.96}$$

if definition (1.91) is used and

$$\mathrm{div}\boldsymbol{T} = \frac{\partial T_{ij}}{\partial x_j}\mathbf{e}_i \tag{1.97}$$

if definition (1.92) is used.

With definition (1.92),

$$\nabla \otimes \boldsymbol{T} = \frac{\partial T_{ij}}{\partial x_k}\boldsymbol{e}_i \otimes \boldsymbol{e}_j \otimes \boldsymbol{e}_k,$$

which is a third-order tensor. When $\partial T_{ij}/\partial x_k$ is contracted with respect to the suffixes i and k, the components of div$\boldsymbol{T}$ defined by equation (1.91) are obtained, and if contracted with respect to j and k, the components defined by equation (1.92) are obtained. The gradient of a tensor of higher order than two with components $\partial S_{ij\ldots}/\partial x_m$ can be contracted with respect to m and any one of the suffixes $i,j,k \ldots l$ or any two of the suffixes $i,j,k \ldots l$.

There is an important relationship between curl $\boldsymbol{v}$ and $\nabla \otimes \boldsymbol{v}$. The symmetric part of $\nabla \otimes \boldsymbol{v}$,

$$\boldsymbol{D} = \frac{1}{2}\left(\nabla \otimes \boldsymbol{v} + (\nabla \otimes \boldsymbol{v})^{\mathrm{T}}\right)$$

has components

$$D_{ij} = \frac{1}{2}\left(v_{i,j} + v_{j,i}\right),$$

where we have adopted the notation $v_{i,j}$ for $\partial v_i/\partial x_j$, and the antisymmetric part

$$\boldsymbol{\Omega} = \frac{1}{2}\left(\nabla \otimes \boldsymbol{v} - (\nabla \otimes \boldsymbol{v})^{\mathrm{T}}\right)$$

has components

$$\boldsymbol{\Omega}_{ij} = \frac{1}{2}\left(v_{i,j} - v_{j,i}\right).$$

Now consider the vector $\boldsymbol{w} = \frac{1}{2}\text{curl } \boldsymbol{v}$, which in component form is,

$$\boldsymbol{w} = \frac{1}{2}\left[\mathbf{e}_1\left(v_{3,2} - v_{2,3}\right) + \mathbf{e}_2\left(v_{1,3} - v_{3,1}\right) + \mathbf{e}_3\left(v_{2,1} - v_{1,2}\right)\right]. \tag{1.98}$$

We note that the independent components of $\boldsymbol{\Omega}_{ij}$ are related to the components of the vector $\boldsymbol{w}$ as follows,

$$\Omega_{ij} = -\mathbf{e}_{ijk} w_k, \tag{1.99}$$

$$w_k = -\frac{1}{2} e_{ijk} \Omega_{ij}. \tag{1.100}$$

It follows from equation (1.99) that $\boldsymbol{w} \times \boldsymbol{a} = \boldsymbol{\Omega a}$ where $\boldsymbol{\Omega} = \boldsymbol{\Omega}_{ij}\mathbf{e}_i \times \mathbf{e}_j$. We can obtain a vector $\boldsymbol{b}$ from any second-order antisymmetric tensor $\boldsymbol{B}$ as follows, $b_i = \mathbf{e}_{ijk} B_{jk}$. Actually, a vector $\boldsymbol{c}$ can be obtained from any second-order tensor $\boldsymbol{C}$ as follows, $c_i = \mathbf{e}_{ijk} C_{jk}$, $c_i = \mathbf{e}_{ijk} C_{jk}$; however, $\boldsymbol{c}$ is the null vector if $\boldsymbol{C}$ is symmetric, and if $\boldsymbol{C}$ is not symmetric, the symmetric part does not influence $\boldsymbol{c}$ since

$$\mathbf{e}_{ijk} C_{jk}^{(s)} = 0, \tag{1.101}$$

where $\boldsymbol{C}^{(s)}$ is the symmetric part of $\boldsymbol{C}$.

The relation

$$\text{tr}(\boldsymbol{AB}) = A_{ji} B_{ij} = 0,$$

where $\boldsymbol{A}$ is symmetric and $\boldsymbol{B}$ is antisymmetric, is similar to equation (1.101).

1.13 Integral Theorems

The integral theorem, which is most frequently encountered in the study of continuum mechanics, is the divergence theorem. In what follows we present the divergence theorem and Stokes' theorem without proof, since proofs of the basic theorem are presented in numerous texts on elementary vector analysis. We consider a bounded regular region of $\boldsymbol{E}_3$, that is, the union of the set of points on a closed piecewise smooth boundary or boundaries and the interior points. It may be simpler to think of a volume bounded by an outer surface, or by an outer surface and one of more inner surfaces. If ϕ is a continuous scalar field, then

$$\int_V \nabla\phi dv = \int_{\partial V} \phi \boldsymbol{n} da,$$

or, in suffix notation,

$$\int_V \frac{\partial \phi}{\partial x_i} dv = \int_{\partial V} \phi \boldsymbol{n}_i da, \tag{1.102}$$

where v denotes the region, ∂v is the boundary with area a, and $\boldsymbol{n}$ is the outward unit normal to the boundary. The divergence theorem

$$\int_v \mathrm{div} u dv = \int_{\partial v} u \cdot n da,$$

or, in suffix notation,

$$\int_v \frac{\partial u_i}{\partial x_i} dv = \int_{\partial v} u_i n_i da \tag{1.103}$$

is obtained from equation (1.97) when ϕ is replaced by the components u_i of a vector $\boldsymbol{u}$, and the summation convention applied. Equation (1.03) is what is usually known as the divergence theorem, although equation (1.02) is more fundamental. The theorem can also be applied to second- and higher-order tensors. For example, if $\boldsymbol{T}$ is a second-order tensor,

$$\int_V \mathrm{div}\boldsymbol{T}d\mathrm{v} = \int_{\partial V} T_{ij}\boldsymbol{n}_j\boldsymbol{da}$$

where div$\boldsymbol{T}$ is given by equation (1.97), so that in suffix notation we have,

$$\int_V \frac{\partial T_{ij}}{\partial x_j} dv = \int_{\partial V} T_{ij} n_j da.$$

EXAMPLE PROBLEM 1.6. Use the divergence theorem to prove that

$$\int_{\partial R} \boldsymbol{n} \times \boldsymbol{u} da = \int_R \mathrm{curl}\boldsymbol{u} dv,$$

where $\boldsymbol{u} \in \mathrm{C}^1$ is a vector field in region R and $\boldsymbol{n}$ is the unit normal to the boundary ∂R of R.

SOLUTION.

$$\int_{\partial R} \boldsymbol{n} \times \boldsymbol{u} da = \left(\int_{\partial R} \mathrm{e}_{ijk} \boldsymbol{n}_j u_k da \right) \mathbf{e}_i = \left(\int_R \mathrm{e}_{ijk} \frac{\partial u_k}{\partial x_j} dv \right) \mathbf{e}_i = \int_R \mathrm{curl}\boldsymbol{u} dv.$$

The second theorem that frequently occurs is Stokes theorem. Let dA be a part of a surface A bounded by a closed curve C, and $\boldsymbol{u}$ be a continuous vector field in the domain containing A, then Stokes theorem is

$$\oint \boldsymbol{u} \cdot dx = \int_A \mathrm{curl}\boldsymbol{u} \cdot \boldsymbol{n} dA$$

where $\oint$ denotes the integral around the curve C, $d\boldsymbol{x}$ is a vector element of C, and $\boldsymbol{n}$ is the unit normal to A taken in the sense of the right-hand rule.

EXERCISES

1.1. Derive the vector relations

(a) $curl\ \nabla\ \alpha = 0.$

(b) $divcurl\ \boldsymbol{u} = 0.$

(c) $div(\alpha \boldsymbol{u}) = \alpha \mathrm{div}\boldsymbol{u} + \boldsymbol{u} \cdot \mathrm{grad}\alpha.$

(d) $curl(\alpha \boldsymbol{u}) = \alpha \mathrm{curl}\boldsymbol{u} + \nabla\alpha \times \boldsymbol{u}.$

(e) $curlcurl\boldsymbol{u} = \nabla \mathrm{div}\boldsymbol{u} - \nabla^2 \boldsymbol{u}.$

(f) $\alpha \nabla^2 \beta = \mathrm{div}(\alpha \nabla \beta) - \nabla \alpha \cdot \nabla \beta.$

(g) $\mathbf{u} \times curl\mathbf{u} = \frac{1}{2}\nabla(\boldsymbol{u} \cdot \boldsymbol{u}) - \boldsymbol{u} \cdot \nabla \boldsymbol{u}.$

1.2. Consider the tensor $\boldsymbol{I} - \boldsymbol{n} \otimes \boldsymbol{n} = (\delta_{ij} - n_i n_j)\mathbf{e}_i \otimes \mathbf{e}_j$, where $\boldsymbol{n}$ is a unit vector. Show that $(\boldsymbol{I} - \boldsymbol{n} \otimes \boldsymbol{n})\boldsymbol{u}$ is the projection of $\boldsymbol{u}$ onto a plane normal to $\boldsymbol{n}$ and that

$$(\boldsymbol{I} - \boldsymbol{n} \otimes \boldsymbol{n})^2 = (\boldsymbol{I} - \boldsymbol{n} \otimes \boldsymbol{n}).$$

1.3. Show that if $\boldsymbol{S}$ is a second-order positive definite tensor

$$\det\left[\boldsymbol{S}^{1/2}\right] = \{\det[\boldsymbol{S}]\}^{1/2}.$$

1.4. Let T^*_{ij} be the components of the deviatoric part of the second-order symmetric tensor $\boldsymbol{T}^*$. Prove that

$$\det[\boldsymbol{T}^*] = \frac{1}{3}\mathrm{tr}(\boldsymbol{T}^*3).$$

1.5. Consider the nine quantities A_{ij} referred to a given coordinate system. If $A_{ij}B_{ij}$ transforms as a scalar under change of coordinate system for an arbitrary second-order symmetric tensor $\boldsymbol{B}$, prove that $A_{ij} + A_{ij}$ are components of a second-order tensor.

1.6. Show that for a proper orthogonal tensor $\boldsymbol{Q}$, there exists a vector $\boldsymbol{u}$ such that $\boldsymbol{Q}\boldsymbol{u} = \mathbf{u}$.

1.7. Show that $\det[\boldsymbol{u} \otimes \boldsymbol{v}] = 0$.

1.8. Consider the positive definite matrix

$$[\boldsymbol{B}] = \begin{bmatrix} 3 & 2 & 1.5 \\ 2 & 4 & 1 \\ 1.5 & 1 & 2 \end{bmatrix}.$$

Obtain the square root,

$$\left[\boldsymbol{B}^{1/2}\right] = \begin{bmatrix} 1.57443 & 0.542316 & 0.476527 \\ 0.542316 & 1.91127 & 0.230106 \\ 0.476527 & 0.230106 & 1.31148 \end{bmatrix}$$

and check this result using the Mathematica command MatrixPower[***B***,1/2].

1.9. Show that for any proper orthogonal tensor $\boldsymbol{Q}$,

$$\det[\boldsymbol{Q} - \boldsymbol{I}] = 0.$$

It is perhaps easier to show this using symbolic notation, noting that $\boldsymbol{Q}\boldsymbol{Q}^{\mathrm{T}} = \boldsymbol{I}$ and the result follows.

1.10. Let V be a region of E_3 with boundary ∂V and outward unit normal $\boldsymbol{n}$. Let $\boldsymbol{\sigma}$ be a Cauchy stress field and $\boldsymbol{u}$ a vector field both of class C^1. Prove that

$$\int_{\partial V} \boldsymbol{u} \otimes \boldsymbol{\sigma}\boldsymbol{n} da = \int_{\mathrm{V}} \{\boldsymbol{u} \otimes \mathrm{div}\boldsymbol{\sigma} + (\nabla \otimes \boldsymbol{u})\boldsymbol{\sigma}\} dv.$$

1.11. Show that

$$\int_{\mathrm{V}} (\nabla \otimes \boldsymbol{u}) dv = \int_{\partial V} (\boldsymbol{n} \otimes \boldsymbol{u}) da,$$

where $\boldsymbol{n}$ is the unit normal to the suface ∂v.
Assume definition (1.91) for ∇.

1.12. If $b = \phi\nabla\psi$, where ϕ and ψ are scalar functions, show that

$$\int_V \left(\phi\nabla^2\psi + \nabla\phi \cdot \nabla\psi\right)dV = \int_{\partial V} \phi\nabla\psi \cdot \boldsymbol{n}dA,$$

where $\boldsymbol{n}$ is the unit normal to the suface A.

SUGGESTED READING

1. Jeffreys, H. (1931). Cartesian Tensors. Cambridge University Press. This was the first text on Cartesian tensor analysis. Does not introduce base vectors or symbolic notation.
2. Temple, G. (1960). Cartesian Tensors. Methuen. Presents a different approach to the definition of a tensor. Does not introduce base vectors or symbolic notation.
3. Prager, W. (1961). Introduction to Mechanics of Continua. Ginn. Introduces base vectors and limited use of symbolic notation. Gives proofs of divergence and Stokes theorems.
4. Chadwick, P. (1976). Continuum Mechanics. George Allen and Unwin. Uses Symbolic notation exclusively. Many worked problems.
5. Gurtin, M.E. (1981). An Introduction to Continuum Mechanics. Academic Press. Uses symbolic notation almost exclusively.
6. Pearson, C.E. (1959). Theoretical Elasticity. Harvard University Press. Has a useful introduction to Cartesian Tensor analysis. Does not introduce base vectors or symbolic notation.
7. Stratton, J.A. (1941). Electro-Magnetic Theory. McGraw-Hill. Has a good introduction to Cartesian Tensor analysis. Introduces base vectors. Has an excellent treatment of orthogonal curvilinear coordinate systems. Does not use summation convention.
8. Leigh, D.C. (1968). Nonlinear Continuum Mechanics. McGraw-Hill. Introduces suffix notation, for both Cartesian and curvilinear coordinates, along with symbolic notation.
9. Lichnerowicz, A. (1962). Tensor Calculus. Methuen. Has an excellent treatment of linear vector spaces and of general tensor analysis.
10. Strang, G. (1993). Introduction to Linear Algebra. Wellesley Cambridge Press. An excellent text which provides the background in linear algebra required for the further study of tensor analysis.

11. Kellogg, O.D. (1953). Foundations of Potential Theory. Dover. This is a classic text, first published in 1929. It contains a detailed discussion of the integral theorems.
12. Flügge, W. (1972). Tensor Analysis and Continuum Mechanics. Springer-Verlag. Contains an excellent treatment of general tensor analysis and its applications to continuum mechanics.

2 Kinematics and Continuity Equation

2.1 Description of Motion

In this chapter we are concerned with the motion of continuous bodies without reference to the forces producing the motion. A continuous body is a hypothetical concept and is a mathematical model for which molecular structure is disregarded and the distribution of matter is assumed to be continuous. Also it may be regarded as an infinite set of particles occupying a region of Euclidean point space $\boldsymbol{E}_3$ at a particular time t. The term particle is used to describe an infinitesimal part of the body, rather than a mass point as in Newtonian mechanics. A particle can be given a label, for example, $\boldsymbol{X}$, and there is a one-one correspondence between the particles and triples of real numbers that are the coordinates at time t with respect to a rectangular Cartesian coordinate system.

There are four common descriptions of the motion of a continuous body:

1. Material description. The independent variables are the particle $\boldsymbol{X}$ and the time t.
2. Referential description. The independent variables are the position vector $\boldsymbol{X}$ of a particle, in some reference configuration, and the time t. The reference configuration could be a configuration that the body never occupies but it is convenient to take it as the actual unstressed undeformed configuration at time $t = 0$. The term natural reference configuration is used to describe the unstressed undeformed configuration at a uniform reference temperature. A difficulty arises when a body has an unloaded residually stressed configuration due perhaps

to lack of fit of component parts. This is not discussed in this chapter. The reference configuration is often called the material, or Lagrangian configuration and the components X_K, $K \in \{1,2,3\}$, with respect to a rectangular Cartesian coordinate system, are the material coordinates of $\boldsymbol{X}$. In many applications, for example, in the theory of elasticity, the reference configuration is taken to be the natural reference configuration.

3. Spatial description. The independent variables are the current position vector $\boldsymbol{x}$ and the present time t. This is the description used in fluid mechanics. The components x_i of $\boldsymbol{x}$ are known as the spatial coordinates and sometimes as the Eulerian coordinates.
4. Relative description. The independent variables are the current position vector $\boldsymbol{x}$ and a variable time $\boldsymbol{\tau}$. The variable time $\boldsymbol{\tau}$ is the time at which the particle occupied another position with position vector ς. The motion is described with ς as the dependent variable, that is,

$$\varsigma = \boldsymbol{X}_t(\boldsymbol{x}, \boldsymbol{\tau}),$$

where the subscript t indicates that the reference configuration is the configuration occupied at time t or that $\boldsymbol{x}$ is the position vector at time t. That is, we have a variable reference configuration.

Descriptions (2) and (3) are emphasized in the applications that follow and (1) and (4) will play only a minor role in the topics considered in this book. We will take the reference configuration for description (2) to be the configuration at time $t = 0$, usually the undeformed state of the body, and will refer the material and spatial coordinates to the same fixed rectangular Cartesian coordinate system.

It is convenient to use lowercase suffixes for the spatial configuration, for example, the position vectors of a particle in the spatial and reference configurations are given by

$$\boldsymbol{x} = x_i \mathbf{e}_i,$$

and

$$\boldsymbol{X} = X_K \boldsymbol{E}_K,$$

respectively. Since the same coordinate system is considered, the unit base vectors $\mathbf{e}_i$ and $\boldsymbol{E}_K$ coincide, that is,

$$(\mathbf{e}_1, \mathbf{e}_2, \mathbf{e}_3) = (\boldsymbol{E}_1, \boldsymbol{E}_2, \boldsymbol{E}_3).$$

The motion of a body may be represented by

$$\boldsymbol{x} = \breve{\boldsymbol{x}}(\boldsymbol{X}, t). \tag{2.1}$$

Equation (2.1), for a particular particle with position vector $\boldsymbol{X}$ at $t = 0$, is the parametric equation with parameter t, of a space curve passing through point $\boldsymbol{X}$, and this curve is the particle path of the particle. A stream line is a curve whose tangent at a fixed time at each of its points is in the direction of the velocity vector.

Although the motion is completely determined by equation (2.1), it is often desirable to consider the state of motion at fixed point in space, described by the velocity field

$$\boldsymbol{v} = \breve{\boldsymbol{v}}(\boldsymbol{x}, \boldsymbol{t}). \tag{2.2}$$

Equation (2.2) represents the velocity at time t of the particle occupying the point with position vector $\boldsymbol{x}$. The stream lines for a fixed time t_0 are given by the solution of the three, in general nonlinear, ordinary differential equations,

$$\frac{dx_i}{ds} = \frac{\breve{v}_i}{|\boldsymbol{v}|}, \; i \in \{1,2,3\},$$

where s is the distance measured along the curve so that $d\boldsymbol{x}/ds$ is the unit vector tangential to the curve. A stream line and particle path coincide in steady flow, that is, if $\partial\breve{\boldsymbol{v}}(\boldsymbol{x},t)\big/\partial t \equiv 0$, but the converse is not necessarily true.

2.2 Spatial and Referential Descriptions

The velocity can also be given as a function of $\boldsymbol{X}$ and t,

$$\boldsymbol{v} = \boldsymbol{V}(\boldsymbol{X}, t). \tag{2.3}$$

Equation (2.3) represents the velocity at time t of the particle that occupied the point $\boldsymbol{X}$ in the reference configuration.

Similarly it follows from equation (2.1) that any quantity f that may be expressed as a function

$$f = \breve{f}(\boldsymbol{x}, t)$$

of the variables $\boldsymbol{x}$ and t may also be expressed as a function

$$\hat{f}(\boldsymbol{X}, t) = \breve{f}(\boldsymbol{x}(\boldsymbol{X}, t), t)$$

of $\boldsymbol{X}$ and t. These are known as the spatial and referential descriptions, respectively. There are two different partial derivatives of f with respect to time, $\partial\breve{f}(\boldsymbol{x}, t)/\partial t$, which is the rate of change of f with respect to time at a fixed point, and $\partial\hat{f}(\boldsymbol{X}, t)/\partial t$, which is the rate of change of f with respect to time of the particle with initial position $\boldsymbol{X}$ and is known as the material derivative. The material derivative can also be expressed as

$$\frac{d\breve{f}}{dt} = \lim_{t\to 0}\frac{1}{\Delta t}\left\{\breve{f}(x + \boldsymbol{v}\Delta t, t + \Delta t) - \breve{f}(\boldsymbol{x}, t)\right\}. \tag{2.4}$$

Using suffix notation and expanding the right-hand side of equation (2.4) as a Taylor series gives[1]

$$\frac{d\breve{f}}{dt} = \lim \frac{1}{\Delta t}\left[\frac{\partial\breve{f}}{\partial x_i} v_i \Delta t + \frac{\partial\breve{f}}{\partial t}\Delta t + O\left(\Delta t^2\right)\right], = \breve{v}_i \frac{\partial\breve{f}}{\partial x_i} + \frac{\partial\breve{f}}{\partial t}, \tag{2.5}$$

[1]The meaning of the O symbol is as follows: $f(x) = O(g(x))\text{as}x \to 0$ implies that $\lim_{x\to 0}|f(x)|/|g(x)| = K$ where K is bounded. When $K = 0$, we use the small o symbol, that is, $f(x) = og(x)$ implies that $\lim_{x\to 0}|f(x)|/|g(x)| = 0$.

or, in symbolic notation,

$$\frac{d\breve{f}}{dt} = \boldsymbol{v} \cdot \nabla\breve{f} + \frac{\partial\breve{f}}{\partial t}. \tag{2.6}$$

Since $\boldsymbol{v} \cdot \nabla$ is a scalar differential operator, equation (2.6) is valid if f is a scalar or tensor of any order. An alternative notation for the material derivative $d()/dt$ is $D()/Dt$. The velocity of a particle is a particular case of a material derivative since

$$\breve{\boldsymbol{v}} = \frac{d\boldsymbol{x}}{dt} = \frac{\partial(\boldsymbol{x}(\boldsymbol{X},t))}{\partial t} = \dot{\boldsymbol{x}}. \tag{2.7}$$

The motion is called steady flow if the velocity is independent of time, that is, if $\partial\boldsymbol{v}/\partial t = 0$.

2.3 Material Surface

A material surface is a surface embedded in a body so that the surface is occupied by the same particles for all time t. In most problems the boundary of a continuous body is a material surface. An exception is when ablation occurs, as in a burning grain of propellant.

A necessary and sufficient condition for a surface $\phi(x_1, x_2, x_3, t) = 0$ to be a material surface is

$$\frac{d\phi}{dt} = \frac{\partial\phi}{\partial t} + \frac{\partial\phi}{\partial x_i} v_i = 0, \tag{2.8}$$

where $\boldsymbol{v}$ is the particle velocity and $d\phi/dt$ is evaluated on $\phi = 0$. The proof is as follows. Suppose a typical point on the surface, $\phi(\boldsymbol{x}, t) = 0$, is moving with velocity $\boldsymbol{u}$, not necessarily equal to $\boldsymbol{v}$, the velocity of the particle instantaneously occupying that point on the surface. The velocity $\boldsymbol{u}$ must satisfy the condition

$$\frac{\partial\phi}{\partial t} + \frac{\partial\phi}{\partial x_i} u_i = 0 \tag{2.9}$$

since

$$\phi(\boldsymbol{x}, t) = 0.$$

The normal component of $\boldsymbol{u}$ on the exterior normal to $\phi = 0$ is

$$u_{(n)} = \boldsymbol{u} \cdot \boldsymbol{n} = \frac{u_s(\partial\phi/\partial x_s)}{\left(\dfrac{\partial\phi}{\partial x_k}\dfrac{\partial\phi}{\partial x_k}\right)^{1/2}},$$

and with the use of equation (2.9) this becomes

$$u_{(n)} = \frac{-(\partial\phi/\partial t)}{\left(\dfrac{\partial\phi}{\partial x_k}\dfrac{\partial\phi}{\partial x_k}\right)^{1/2}}. \tag{2.10}$$

The normal component of the particle velocity $\boldsymbol{v}$ on the exterior normal to $\phi = 0$ is

$$v_{(n)} = \frac{v_s(\partial\phi/\partial x_s)}{\left(\dfrac{\partial\phi}{\partial x_k}\dfrac{\partial\phi}{\partial x_k}\right)^{1/2}}. \tag{2.11}$$

Substituting for $\partial\phi/\partial t$ from equation (2.10) and for $v_s(\partial\phi/\partial x_s)$ from equation (2.11) in

$$\frac{d\phi}{dt} = \frac{\partial\phi}{\partial t} + \frac{\partial\phi}{\partial x_s} v_s$$

gives

$$\frac{d\phi}{dt} = \left|v_{(n)} - u_{(n)}\right| \left[\frac{\partial\phi}{\partial x_k}\frac{\partial\phi}{\partial x_k}\right]^{1/2}. \tag{2.12}$$

If the surface always consists of the same particles, $u_{(n)} = v_{(n)}$, and it follows from equation (2.12) that $d\phi/dt = 0$, evaluated on $\phi = 0$, is a necessary condition for $\phi = 0$ to be a material surface. This proof of the necessity of the condition can be given in words. If the surface is a material surface, the velocity, relative to the surface, of a particle lying in it must be wholly tangential (or zero); otherwise we would have flow across the surface. It follows that the instantaneous rate of variation of ϕ for a surface particle must be zero, if $\phi(\boldsymbol{x}, t) = 0$ is a material surface. To prove the sufficiency of the condition we put

$$\Phi(\boldsymbol{X}, t) = \phi(\boldsymbol{x}(\boldsymbol{X}, t), t) = 0, \tag{2.13}$$

so that $d\phi/dt = 0$ implies that

$$\frac{\partial \Phi(\boldsymbol{X}, t)}{\partial t} = 0.$$

This further implies that $\Phi = 0$ is the equation of a surface fixed in the reference configuration, and it then follows from equation (2.13) that $\phi(\boldsymbol{x}, t)$ always contains the same particles.

2.4 Jacobian of Transformation and Deformation Gradient *F*

Let dV be the volume of an infinitesimal material parallelepiped element of a continuous body in the reference configuration, taken as the undeformed configuration, and let dv be the volume of the parallelepiped in the current or spatial configuration. Contiguous sides of the parallelepiped are the material line elements $d\boldsymbol{X}^{(1)}$, $d\boldsymbol{X}^{(2)}$, $d\boldsymbol{X}^{(3)}$ in the reference configuration, and $d\boldsymbol{x}^{(1)}$, $d\boldsymbol{x}^{(2)}$, $d\boldsymbol{x}^{(3)}$ in the spatial configurations are as shown in Figure 2.1.

The volume of a parallelepiped is given by the triple scalar product of the vectors, representing contiguous sides, taken in the appropriate order, consequently,

$$dV = d\mathbf{X}^{(1)} \times d\mathbf{X}^{(2)} \cdot d\mathbf{X}^{(3)} = e_{KLM} dX_L^{(1)} dX_M^{(2)} dX_K^{(3)}, \tag{2.14}$$

$$dv = dx^{(1)} \times dx^{(2)} \cdot dx^{(3)} = \mathrm{e}_{ijk} dx_i^1 dx_j^2 dx_k^3. \tag{2.15}$$

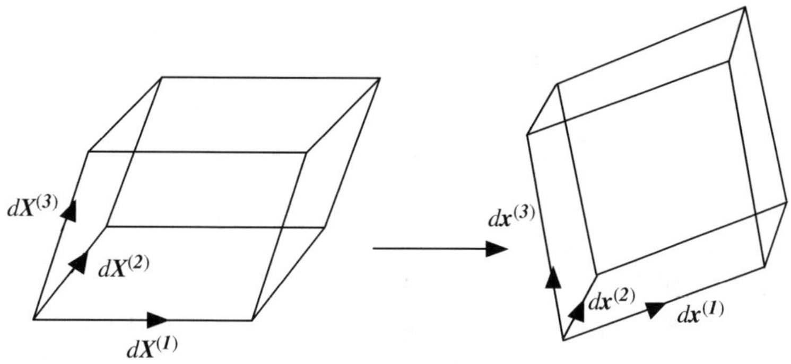

Figure 2.1. Deformation of infinitesimal parallelepiped.

Substituting

$$dx_i^{(s)} = \frac{\partial x_i}{\partial X_L} dX_L^{(s)} + O\left(\left|dX^{(s)}\right|^2\right),\ s \in \{1,2,3\} \tag{2.16}$$

in equation (2.15) gives

$$dv = \mathrm{e}_{ijk} \frac{\partial x_i}{\partial X_K} \frac{\partial x_j}{\partial X_L} \frac{\partial x_k}{\partial X_M} dX_L^{(1)} dX_M^{(2)} dX_K^{(3)}, \tag{2.17}$$

neglecting terms $O\left(\left|d\boldsymbol{X}^{(s)}\right|^2\right)$.

It follows from equations (1.71), (2.14), and (2.17) that

$$dv = JdV, \tag{2.18}$$

where

$$J = \det\left[\frac{\partial x_i}{\partial X_K}\right], \tag{2.19}$$

or, using a common notation,

$$J = \frac{\partial(x_1, x_2, x_3)}{\partial(X_1, X_2, X_3)}.$$

The determinant J is the Jacobian of the transformation (2.1) and is sometimes known as the dilatation because of equation (2.18). It is evident from physical considerations that

$$0 < J < \infty,$$

and this also follows from the invertibility of equation (2.1). A motion for which there is no volume change for any element of the body is described as isochoric. If a body is assumed to be incompressible, it can only undergo isochoric motions, otherwise thermodymamic anomalies arise. It follows from equation (2.18) that $J = 1$ for an isochoric motion.

The elements $F_{ik} = \partial x_i / \partial X_K$ of the matrix in equation (2.19) are the components of a second-order two-point tensor,

$$\boldsymbol{F} = F_{iK} \mathbf{e}_i \otimes \mathbf{E}_K, \tag{2.20}$$

known as the deformation gradient tensor. It is implied by equation (2.20) that $\boldsymbol{F}$ transforms as a vector for coordinate transformations in one configuration. This shows that $\boldsymbol{F}$ is related to both configurations and it is a linear operator that operates on a vector in the reference configuration to give a vector in the spatial configuration. If $(\boldsymbol{X}, \boldsymbol{x})$ and $(\boldsymbol{X}^{\boldsymbol{O}}, \boldsymbol{x}^{\boldsymbol{O}})$ are the position vectors of two adjacent particles in the reference and spatial configurations, respectively, then

$$x_i - x_i^O = \frac{\partial x_i}{\partial X_K} \left(X_K - X_K^O \right) + \mathrm{O}\left(\left| X_K - X_K^O \right|^2 \right). \tag{2.21}$$

When $\boldsymbol{X}$ and $\boldsymbol{X}^O$ are an infinitesimal distance apart we can put $d\boldsymbol{X} = \boldsymbol{X} - \boldsymbol{X}^O$ and $d\boldsymbol{x} = \boldsymbol{x} - \boldsymbol{x}^O$ so that

$$d\boldsymbol{x} = \boldsymbol{F} d\boldsymbol{X}, dx_i = F_{iK} dX_K. \tag{2.22}$$

Equation (2.21) should be compared with equation (2.16). The term two-point tensor indicates the relationship to both the spatial and reference configurations.

The deformation gradient tensor $\boldsymbol{F}$ can also be expressed in the form

$$\boldsymbol{F} = \text{Grad}\boldsymbol{x}$$

where

$$\text{Grad} \equiv \frac{\partial()}{\partial X_K}\boldsymbol{E}_K.$$

The inverse $\boldsymbol{F}^{-1}$ of the deformation gradient tensor occurs frequently, and its components are given by

$$F_{Ki}^{-1} = \frac{\partial X_K}{\partial x_i}, \boldsymbol{F}^{-1} = \frac{\partial X_K}{\partial x_i}\boldsymbol{E}_K \otimes \mathbf{e}_i \tag{2.23}$$

and the inverse of equation (2.22) is

$$d\boldsymbol{X} = \boldsymbol{F}^{-1} d\boldsymbol{x}.$$

2.5 Reynolds Transport Theorem

We have already considered a material surface that is a surface that contains the same particles for all time. Material lines and material volumes are similar concepts. A material volume is bounded by a material surface or surfaces and is frequently referred to as a system volume, especially in the literature of thermodynamics.

There are certain theorems concerning the material derivatives of line, surface, and volume integrals of a field $\phi(\boldsymbol{x}, t)$ where ϕ may be a scalar, vector, or higher-order tensor. The most important of these theorems is that for the volume integral that is usually known as Reynolds Transport Theorem. This theorem can be expressed in the form

$$\frac{d}{dt}\int_{v(t)} \phi(\boldsymbol{x}, t) d\mathrm{v} = \int_{v(t)} \frac{\partial \phi}{\partial t} d\mathrm{v} + \int_{\partial v(t)} \phi \boldsymbol{v} \cdot \boldsymbol{n} dS, \tag{2.24}$$

where $\boldsymbol{v}$ is the velocity field, $\boldsymbol{n}$ is the outward unit normal to the surface S of the material volume $\mathrm{v}(t)$. According to equation (2.24), the rate of change of the integral of $\boldsymbol{\phi}$ over the material volume $\mathrm{v}(t)$ is equal to the integral of $\partial\phi/\partial t$ over the fixed volume that coincides instantaneously with $\mathrm{v}(t)$ at time t plus the flux $\phi\boldsymbol{v}$ through the surface S of this fixed volume.

In order to prove equation (2.24) we need the result

$$\frac{dJ}{dt} = J\nabla \cdot \boldsymbol{v}, \tag{2.25}$$

which is due to Euler's Equation. The proof of equation (2.25) is as follows. First we note that the Jacobian J is given by

$$J\delta_{ij} = \frac{\partial x_i}{\partial X_K}\mathbf{A}_{jK}, \tag{2.26}$$

where $\boldsymbol{A}_{iK}$ is the cofactor of $F_{iK} = \partial x_i/\partial X_K$ in the determinant J. Since the derivative of J with respect to t is the sum of the three determinants formed by replacing the elements of a different row (or column) of each determinant by their time derivatives,

$$\frac{dJ}{dt} = \frac{d}{dt}\left(\frac{\partial x_i}{\partial X_K}\right)A_{iK} = \frac{\partial v_i}{\partial X_K}A_{iK} = \frac{\partial v_i}{\partial x_j}\frac{\partial x_j}{\partial X_K}A_{iK},$$

and with the use of equation (2.26) it follows that

$$\frac{dJ}{dt} = J\frac{\partial v_i}{\partial x_i} = J\nabla \cdot \boldsymbol{v}. \tag{2.27}$$

Using equations (2.1) and (2.18) we obtain

$$\frac{d}{dt}\int\limits_{v(t)} \phi(\boldsymbol{x}, t)d\mathrm{v} = \frac{\partial}{\partial t}\int\limits_{V} \left(J\hat{\phi}(\boldsymbol{X}, t)\right)dV, \tag{2.28}$$

where $\hat{\phi}(\boldsymbol{X}, t) = \phi(\boldsymbol{x}(\boldsymbol{X}, t), t)$. Since the integral on the right-hand side of equation (2.28) is over a fixed volume we can differentiate inside the integral sign to obtain

$$\frac{d}{dt}\int_{v(t)} \phi(\boldsymbol{x}, t)\mathrm{dv} = \int_{V}\left(J\frac{\partial\hat{\phi}}{\partial t} + \hat{\phi}\frac{dJ}{dt}\right)dV,$$

and with the use of equation (2.27) this becomes

$$\frac{d}{dt}\int_{v(t)} \phi(\boldsymbol{x}, t)dv = \int_{v(t)}\left(\frac{d\phi}{dt} + \phi\mathrm{div}\boldsymbol{v}\right)dv. \tag{2.29}$$

Equation (2.24) can be obtained from equation (2.29) with the use of the identity

$$\nabla\cdot(\phi\boldsymbol{v}) = \phi\nabla\cdot\boldsymbol{v} + \boldsymbol{v}\cdot\nabla\phi,$$

relation (2.6), with $\breve{f}$ replaced by ϕ, and the divergence theorem.

Equations (2.24) and (2.29) are valid when $\phi(\boldsymbol{x}, t)$ is continuous. Suppose $\phi(\boldsymbol{x}, t)$ has a discontinuity across a surface $\sum(t)$, that moves with velocity $\boldsymbol{w}(\boldsymbol{x}, t)$, not necessarily equal to the field velocity $\boldsymbol{v}$, and divides the volume v into two parts, v^+ and v^-, as indicated in Figure 2.2. The unit normal $\boldsymbol{m}$ to $\sum(t)$ points from the –side to the +side.
Then, equations (2.24) and (2.29) must be replaced by equivalent forms

$$\frac{d}{dt}\int_{v(t)} \phi(\boldsymbol{x}, t)\mathrm{dv} = \int_{v(t)}\frac{\partial\phi}{\partial t}\mathrm{dv} + \int_{\partial v(t)} \phi\boldsymbol{v}\cdot\boldsymbol{n}dS - \int_{\Sigma(t)} [\![(\boldsymbol{v}-\boldsymbol{w})\cdot\boldsymbol{m}\phi]\!]\mathrm{da}, \tag{2.30a}$$

$$\frac{d}{dt}\int_{v(t)} \phi(\boldsymbol{x}, t)\mathrm{dv} = \int_{v(t)}\left(\frac{d\phi}{dt} + \phi\mathrm{div}\boldsymbol{v}\right)\mathrm{dv} - \int_{\Sigma(t)} [\![(\boldsymbol{v}-\boldsymbol{w})\cdot\boldsymbol{m}\phi]\!]da, \tag{2.30b}$$

respectively, where $[\![(\boldsymbol{v}-\boldsymbol{w})\cdot\boldsymbol{m}\phi]\!] = ((\boldsymbol{v}-\boldsymbol{w})\cdot\boldsymbol{m}\phi)^+ - ((\boldsymbol{v}-\boldsymbol{w})\cdot\boldsymbol{m}\phi)^-$ and the superscripts + and – denote the values on the + and –side of $\sum(t)$, respectively. A proof of equation (2.30a) is as follows. The rates of

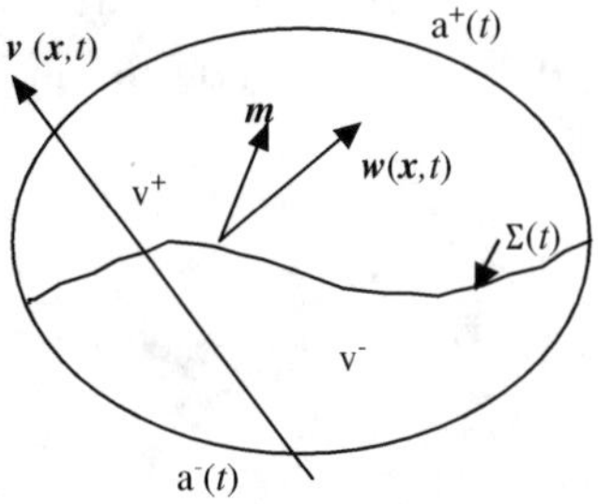

Figure 2.2. Discontinuity across surface $\Sigma(t)$.

change of the integral of ϕ over the volumes $v^+(t)$ and $v^-(t)$are equal to the integral of $\partial\phi/\partial t$ over the fixed volumes that coincide instantaneously with $v^+(t)$ and $v^-(t)$at time t plus the flux $\phi\boldsymbol{v}$ through the surfaces of the fixed volumes. It then follows that

$$\frac{d}{dt}\int_{v^+(t)} \phi(\boldsymbol{x}, t)d\mathrm{v} = \int_{v^+(t)} \frac{\partial\phi}{\partial t}d\mathrm{v} + \int_{a^+(t)} \phi\boldsymbol{v}\cdot\boldsymbol{n}d\mathrm{a} - \int_{\Sigma(\tau)} ((\boldsymbol{v}-\boldsymbol{w})\cdot\boldsymbol{m}\phi^+)d\mathrm{a}$$

$$\frac{d}{dt}\int_{v^-(t)} \phi(\boldsymbol{x}, t)d\mathrm{v} = \int_{v^-(t)} \frac{\partial\phi}{\partial t}d\mathrm{v} + \int_{a^-(t)} \phi v\cdot\boldsymbol{n}d\mathrm{a} + \int_{\Sigma(\tau)} ((\boldsymbol{v}-\boldsymbol{w})\cdot\boldsymbol{m}\phi^-)d\mathrm{a},$$

and addition of these equations gives equation (2.30a)

Equation (2.30) has applications in wave propagation problems that involve shock waves.

The one-dimensional form of equation (2.24) is,

$$\frac{d}{dt}\int_{x_1(t)}^{x^2(t)} \phi(x, t)dx = \int_{x_1(t)}^{x^2(t)} \frac{\partial\phi}{\partial t}dx + \dot{x}_2\phi(x_2, t) - \dot{x}_1\phi(x_1, t),$$

where a superposed dot denotes differentiation with respect to t. This is just Leibniz's rule in elementary calculus for differentiation of an integral with variable limits.

The theorems that involve material differentiation of line and surface integrals are given in the problem set at the end of this chapter.

2.6 Continuity Equation

Conservation of mass requires that

$$\frac{d}{dt}\int_{\mathrm{v}(t)} \rho(\boldsymbol{x}, t)d\mathrm{v} = 0,$$

where ρ is the density in the spatial configuration and $\mathrm{v}(t)$ is a material volume. It then follows from the form (2.29) of the transport equation that

$$\int_{\mathrm{v}(t)} \left[\frac{d\rho}{dt} + \rho\nabla \cdot \boldsymbol{v}\right] d\mathrm{v} = 0.$$

Since this result must hold for an arbitrary material volume and the integrand is continuous,

$$\frac{d\rho}{dt} + \rho\nabla \cdot \boldsymbol{v} = 0, \tag{2.31}$$

which is the spatial form of continuity equation and which can also be expressed in the form

$$\frac{\partial\rho}{\partial t} + \nabla \cdot (\rho\boldsymbol{v}) = 0,$$

by using the expression for the material derivative of ρ and the relation $\nabla \cdot (\rho\boldsymbol{v}) = \rho\nabla \cdot \boldsymbol{v} + \boldsymbol{v} \cdot \nabla\rho$. The material form of the continuity equation is

$$\rho_O = \rho J, \tag{2.32}$$

where ρ_O is the density in the reference configuration. Equation (2.32) is obtained from equation (2.18) and $\rho_O dV = \rho dv$, which follows from conservation of mass. Equation (2.31) can also be obtained by taking the material derivative of equation (2.32) and using equation (2.22).

A useful result

$$\frac{d}{dt}\int\limits_{v(t)} \rho\varphi dv = \int\limits_{v(t)} \rho\frac{d\varphi}{dt}dv \tag{2.33}$$

can be obtained from equations (2.29) and (2.31).

2.7 Velocity Gradient

The velocity gradient tensor, $\nabla \otimes \boldsymbol{v}$, is the gradient of the velocity vector $\boldsymbol{v}$; consequently, it may be expressed in terms of its components $v_{i,j}$ relative to a rectangular Cartesian coordinate system as

$$\nabla \otimes \boldsymbol{v} = \frac{\partial v_i}{\partial x_j}\mathbf{e}_i \otimes \mathbf{e}_j.$$

A common notation is

$$\boldsymbol{L} = \nabla \otimes \boldsymbol{v},$$

and henceforth, $\boldsymbol{L}$ denotes the velocity gradient.

Material differentiation of the deformation gradient gives

$$\frac{d}{dt}\left[\frac{\partial x_i}{\partial X_K}\right] = \frac{\partial v_i}{\partial X_K} = \frac{\partial v_i}{\partial x_j}\frac{\partial x_j}{\partial X_K},$$

and, in symbolic notation,

$$\dot{\boldsymbol{F}} = \boldsymbol{L}\boldsymbol{F} \tag{2.34}$$

or

$$\boldsymbol{L} = \dot{\boldsymbol{F}}\boldsymbol{F}^{-1}, \tag{2.35}$$

which gives the relation between the velocity gradient and the deformation gradient.

Consider a material line element $\bar{P}\bar{Q}$, and let the velocities of P and Q be $\boldsymbol{v}_O$ and $\boldsymbol{v}$, respectively. It then follows from a Taylor series expansion that the velocity of Q relative to P is

$$\boldsymbol{v} - \boldsymbol{v}_O = \nabla \otimes \boldsymbol{v}(\boldsymbol{x} - \boldsymbol{x}_O) + O\left(|\boldsymbol{x} - \boldsymbol{x}_O|^2\right),$$

where $\boldsymbol{x}_O$ and $\boldsymbol{x}$ are the position vectors of P and Q, respectively. When PQ is an infinitesimal line element we can replace this expression by

$$d\boldsymbol{v} = \nabla \otimes \boldsymbol{v} d\boldsymbol{x}, \tag{2.36}$$

or, in suffix notation,

$$dv_i = \frac{\partial v_i}{\partial x_j} dx_j.$$

We must note the meanings of $d\boldsymbol{x}$ and $d\boldsymbol{v}$, namely, that $d\boldsymbol{x}$ is the vector of the infinitesimal directed line segment $\bar{P}\bar{Q}$ and d$\boldsymbol{v}$ is the velocity of Q relative to P. The material time derivative of $d\boldsymbol{x}$ is $d\boldsymbol{v}$, since

$$\frac{d}{dt}(dx_i) = \frac{d}{dt}\left(\frac{\partial x_i}{\partial X_K} dX_K\right) = \frac{\partial v_i}{\partial X_K} dX_K = \frac{\partial v_i}{\partial X_K}\frac{\partial X_K}{\partial x_j} dx_j = \frac{\partial v_i}{\partial x_j} dx_j = dv_i. \tag{2.37}$$

2.8 Rate of Deformation and Spin Tensors

The symmetric and antisymmetric parts $\boldsymbol{D}$ and $\boldsymbol{W}$, respectively, of the velocity gradient tensor, $\boldsymbol{L} = \nabla \otimes \boldsymbol{v} = \boldsymbol{D} + \boldsymbol{W}$, are known as the rate of deformation and spin tensors, respectively, so that

$$\boldsymbol{D} = \frac{1}{2}\left(\boldsymbol{L} + \boldsymbol{L}^T\right) \tag{2.38}$$

and

$$\boldsymbol{W} = \frac{1}{2}\left(\boldsymbol{L} - \boldsymbol{L}^T\right). \tag{2.39}$$

We use notations d_{ij} and w_{ij} for the components of $\boldsymbol{D}$ and $\boldsymbol{W}$, respectively, so that

$$d_{ij} = \frac{1}{2}\left(\frac{\partial v_i}{\partial x_j} + \frac{\partial v_j}{\partial x_i}\right)$$

and

$$w_{ij} = \frac{1}{2}\left(\frac{\partial v_i}{\partial x_j} - \frac{\partial v_j}{\partial x_i}\right).$$

The material derivative of the square of the length ds of an infinitesimal material line element $\bar{P}\bar{Q} = d\boldsymbol{x}$ is

$$\begin{aligned}\frac{d}{dt}\left(ds^2\right) &= \frac{d}{dt}(dx_i dx_i) = 2dx_i \frac{d}{dt}(dx_i) = 2\frac{\partial v_i}{\partial x_j} dx_j dx_i = 2\left(d_{ij} + \omega_{ij}\right) dx_i dx_j \\ &= 2d_{ij} dx_i dx_j,\end{aligned} \tag{2.40}$$

since

$$\omega_{ij} dx_i dx_j = 0. \tag{2.41}$$

Also note that

$$\frac{d}{dt}\left(ds^2\right) = 2ds \frac{d}{dt}(ds),$$

and using this along with equation (2.40) gives

$$\frac{1}{ds}\frac{d}{dt}(ds) = d_{ij} \frac{dx_i}{ds}\frac{dx_j}{ds} = d_{ij} n_i n_j. \tag{2.42}$$

Equation (2.42) gives an insight into the physical significance of the rate of deformation tensor $\boldsymbol{D}$, which is often called the strain rate tensor especially in plasticity theory.

It follows from equation (2.40) or (2.42) that a necessary and sufficient condition for a rigid body motion, that is, a motion for which the distance between any two particles remains invariant, is $d_{ij} = 0$. This is known as Killing's theorem. It was shown in the previous section that the velocity of Q relative to P is $d\boldsymbol{v} = \nabla \otimes \boldsymbol{v} d\boldsymbol{x}$ for the material line element $\bar{P}\bar{Q} = d\boldsymbol{x}$, and it may be deduced from Killing's theorem that the part $\boldsymbol{W} d\boldsymbol{x}$ of $\nabla \otimes \boldsymbol{v} d\boldsymbol{x} = \boldsymbol{L} d\boldsymbol{x}$ corresponds to a rigid body motion.

In order to see the significance of the spin tensor $\boldsymbol{W}$, it is desirable to introduce the vector

$$\boldsymbol{\omega} = \frac{1}{2}\text{curl}\boldsymbol{v} = \frac{1}{2}\left\{\mathbf{e}_1\left(\frac{\partial v_3}{\partial x_2} - \frac{\partial v_2}{\partial x_3}\right) + \mathbf{e}_2\left(\frac{\partial v_1}{\partial x_3} - \frac{\partial v_3}{\partial x_1}\right) + \mathbf{e}_3\left(\frac{\partial v_2}{\partial x_1} - \frac{\partial v_1}{\partial x_2}\right)\right\}.$$

In fluid mechanics the vector $\varsigma = 2\omega = \text{curl}\boldsymbol{v}$ is known as the vorticity vector.

Comparison of the components of the vorticity vector 2ω with the components of the spin tensor shows that ω is the vector of the antisymmetric spin tensor $\boldsymbol{W}$. Consequently,

$$\omega_i = -\frac{1}{2} e_{ijk} w_{jk}, w_{ij} = e_{ijk}\omega_k. \tag{2.43}$$

The significance of the vector ω and the tensor $\boldsymbol{W}$ then becomes evident from the following theorem. The angular velocity of the triad of material line elements that instantaneously coincides with the principal axes of the rate of deformation tensor $\boldsymbol{D}$ is given by $\boldsymbol{\omega} = \frac{1}{2}\text{curl}\boldsymbol{v}$. In order to prove this theorem we need to show that

$$\frac{d}{dt}\boldsymbol{n}^{(\alpha)} = \omega \times \boldsymbol{n}^{(\alpha)}, \tag{2.44}$$

where $\boldsymbol{n}^{(\alpha)}, \alpha \in \{1,2,3\}$ are the principal directions or eigenvectors of $\boldsymbol{D}$. This can be shown as follows:

$$\frac{dn_k^{(\alpha)}}{dt} = \frac{d}{dt}\left[\frac{dx_k}{ds^{(\alpha)}}\right] = \frac{1}{ds^{(\alpha)}}\frac{d}{dt}(dx_k) - \frac{1}{ds^{(\alpha)^2}}\frac{d}{dt}\left(ds^{(\alpha)}\right)dx_k, \tag{2.45}$$

where $ds^{(\alpha)}$ is an element of length in the corresponding principal direction. Using equations (2.37), (2.42), (2.43), and

$$\left(d_{ik} - d^{(\alpha)}\delta_{ki}\right)n_i^{(\alpha)} = 0,$$

where $d^{(\alpha)}$ are principal values of the rate of deformation tensor $\boldsymbol{D}$, it follows from equation (2.45) that

$$\frac{dn_k^{(\alpha)}}{dt} = \frac{\partial v_k}{\partial x_i}n_i^{(\alpha)} - d^{(\alpha)}n_k^{(\alpha)} = w_{ki}n_i^{(\alpha)} = \mathbf{e}_{kji}\omega_j n_i^{(\alpha)} = (\boldsymbol{\omega}\times\boldsymbol{n})_k,$$

and the result (2.44) is proved.

Another interpretation of the vorticity $2\boldsymbol{\omega}$ is as follows. If an element of a continuum with its mass centre at P is instantaneously made rigid without change of angular momentum, then its angular velocity immediately after it has been made rigid is $\boldsymbol{\omega} = \frac{1}{2}\text{curl}\boldsymbol{v}$, if and only if the principal axes of inertia for the resulting rigid element lie along the principal axes of the rate of deformation tensor evaluated at P immediately before the element becomes rigid. The proof of this depends on certain results in rigid body dynamics and is as follows. Let $\boldsymbol{r}$ be the position vector of a particle of the element with respect to P. The element is assumed to be sufficiently small that terms $O\left(|r|^2\right)$ may be neglected. At the instant immediately before becoming rigid the angular momentum $\boldsymbol{H}$ of the element about its mass center is

$$\boldsymbol{H} = \int_v \rho\boldsymbol{r}\times\boldsymbol{v}^* dv,$$

where v is the volume of the element and the velocity relative to P of a particle, of the element, is given by

$$v_i^* = \frac{\partial v_i}{\partial x_s} r_s + \mathrm{O}\left(|\boldsymbol{r}|^2\right),$$

and $\partial v_i / \partial x_s$ is evaluated at P. The angular momentum $\boldsymbol{H}$ of the element is then given by

$$H_i = \mathrm{e}_{ijk} \frac{\partial v_k}{\partial x_s} \int_{\mathrm{v}} \rho r_j r_s d\mathrm{v} = \mathrm{e}_{ijk} (d_{ks} + w_{ks}) \int_{v} \rho r_j r_s d\mathrm{v}. \tag{2.46}$$

At the instant immediately after becoming rigid without change in angular momentum,

$$H_i = I_{ij} \omega'_j. \tag{2.47}$$

where $\boldsymbol{\omega}'$ is the angular velocity of the rigid element and

$$I_{ij} = \int_{v} \rho \left(r^2 \delta_{ij} - r_i r_j \right) d\mathrm{v}, \tag{2.48}$$

where I_{ij} are the components of the inertia tensor of the element. Also at the instant immediately after the element becomes rigid, the rate of deformation tensor is zero, so that

$$H_i = \boldsymbol{e}_{ijk} \omega'_{ks} \int_{v} \rho r_s r_j d\mathrm{v}, \tag{2.49}$$

where ω'_{ks} are the components of the spin tensor. Equation (2.49) is equivalent to equation (2.47), since substituting equation $(2.38)_2$ in equation (2.43) gives

$$H_i = -\boldsymbol{e}_{ijk}\boldsymbol{e}_{stk}\omega'_t \int_V \rho r_s r_j dv = -(\delta_{ij}\delta_{jt} - \delta_{it}\delta_{js})\omega'_t \int_v \rho r_s r_j dv$$
$$= \omega'_j \int_v \rho\left(\boldsymbol{r}^2\delta_{ij} - r_i r_j\right)dv.$$

Referring to equation (2.46), $d_{ks} \int_v r_s r_j dv$ are the components of a second-order symmetric tensor if and only if d_{rs} and $\int_v \rho r_s r_j dv$ are coaxial, that is, if and only if they have the same principal axes or eigenvectors. Also the principal axes of I_{ij} and $\int_v \rho r_s r_j dv$ coincide. Comparison of equations (2.46) and (2.49) shows that if and only if d_{ij} and I_{ij} are coaxial,

$$w_{ij} = w'_{ij}$$

since

$$\boldsymbol{e}_{ijk} d_{ks} \int_V \rho r_s r_j dv = 0,$$

if and only if $d_{rs} \int_v \rho r_s r_j dv$ is symmetric.

2.9 Polar Decomposition of *F*

The polar decomposition theorem given in chapter 1 can be applied to the deformation gradient tensor $\boldsymbol{F}$ since $\det[\boldsymbol{F}] > 0$, so that we have the unique decompositions,

$$\boldsymbol{F} = \boldsymbol{RU},\ \boldsymbol{F} = \boldsymbol{VR}, \tag{2.50}$$

or, in suffix notation,

$$F_{iK} = R_{iK} U_{KL} = V_{ij} R_{jK}.$$

It follows from equation (2.50) that

$$\boldsymbol{U} = \boldsymbol{R}^T \boldsymbol{V}\boldsymbol{R}, \boldsymbol{V} = \boldsymbol{R}\boldsymbol{U}\boldsymbol{R}^T. \tag{2.51}$$

According to the Polar Decomposition Theorem given in chapter 1, $\boldsymbol{U}$ and $\boldsymbol{V}$ are symmetric positive definite second-order tensors with the same eigenvalues, $\lambda_\alpha, \alpha \in \{1, 2, 3\}$, that are known as the principal stretches, and $\boldsymbol{R}$ is a proper orthogonal tensor so that

$$\det[\boldsymbol{R}] = 1 \text{ and } \boldsymbol{R}^{\mathrm{T}} = \boldsymbol{R}^{-1}.$$

If $\boldsymbol{R} = \boldsymbol{I}$, $\boldsymbol{F} = \boldsymbol{U} = \boldsymbol{V}$, and the deformation is pure strain with no rotation.

The spectral forms of $\boldsymbol{U}$ and $\boldsymbol{V}$ are

$$\boldsymbol{U} = \sum_{\alpha=1}^{3} \lambda_\alpha \boldsymbol{N}^{(\alpha)} \otimes \boldsymbol{N}^{(\alpha)}, \boldsymbol{V} = \sum_{\alpha=1}^{3} \lambda_\alpha \boldsymbol{n}^{(\alpha)} \otimes \boldsymbol{n}^{(\alpha)}, \tag{2.52a}$$

if the λ_α are distinct, where $\boldsymbol{N}^{(\alpha)}$ and $\boldsymbol{n}^{(\alpha)}$ are the eigenvectors of $\boldsymbol{U}$ and $\boldsymbol{V}$, respectively, and

$$\begin{aligned} \boldsymbol{U} &= \lambda_3 \boldsymbol{N}^{(3)} \otimes \boldsymbol{N}^{(3)} + \lambda\left(\boldsymbol{I} - \boldsymbol{N}^{(3)} \otimes \boldsymbol{N}^{(3)}\right), \\ \boldsymbol{V} &= \lambda_3 \boldsymbol{n}^{(3)} \otimes \boldsymbol{n}^{(3)} + \lambda\left(\boldsymbol{I} - \boldsymbol{n}^{(3)} \otimes \boldsymbol{n}^{(3)}\right), \end{aligned} \tag{2.52b}$$

if $\lambda_1 = \lambda_2 = \lambda \neq \lambda_3$, and

$$\boldsymbol{U} = \lambda \boldsymbol{I}, \ \boldsymbol{V} = \lambda \boldsymbol{I}. \tag{2.52c}$$

It follows from equations (2.51) and (2.52a) that

$$\boldsymbol{n}^{(\alpha)} = \boldsymbol{R}\boldsymbol{N}^{(\alpha)}, \ \boldsymbol{N}^{(\alpha)} = \boldsymbol{R}^{\mathrm{T}}\boldsymbol{n}^{(\alpha)}.$$

Since $\boldsymbol{R}$ is an orthogonal tensor, a vector is unchanged in magnitude when operated on by $\boldsymbol{R}$. It follows that for a rigid body motion $\boldsymbol{F} = \boldsymbol{R}$ and $\boldsymbol{U} = \boldsymbol{I}$, $\boldsymbol{V} = \boldsymbol{I}$. This may be shown otherwise since, for a rigid body motion,

$$d\boldsymbol{x} \cdot d\boldsymbol{x} = d\boldsymbol{X} \cdot d\boldsymbol{X}$$

for all material line elements at a point, and using equation (2.20) and the substitution operator property of the Kronecker Delta we obtain

$$F_{iK}F_{iL} = \delta_{KL}$$

for rigid body motion, which implies $\boldsymbol{F} = \boldsymbol{R}$ and $\boldsymbol{U} = \boldsymbol{I}$, $\boldsymbol{V} = \boldsymbol{I}$.

The principal components or eigenvalues, $\lambda_\alpha, \alpha = \{1, 2, 3\}$, of $\boldsymbol{U}$ and $\boldsymbol{V}$ are called the principal stretches of $\boldsymbol{U}$ and $\boldsymbol{V}$, the right and left stretch tensor, respectively. The form $(2.50)_1$ corresponds to a pure deformation followed by a rigid body rotation given by $\boldsymbol{R}$ that is sometimes called the rotation tensor. Similarly the form $(2.50)_2$ corresponds to a rigid body rotation, given by $\boldsymbol{R}$, followed by a pure deformation. A pure deformation is as follows. Consider a material rectangular parallelipiped with sides of length $l_O^{(\alpha)}$, $\alpha = \{1, 2, 3\}$ that is uniformly deformed into a rectangular parallelipiped with corresponding sides in the same directions and of length $l^{(\alpha)}$. Each side undergoes a stretch $\lambda_\alpha = l^{(\alpha)}/l_O^{(\alpha)}$.

As an example of polar decomposition we consider simple shear deformation given by

$$x_1 = X_1 + KX_2,\ x_2 = X_2,\ x_3 = X_3, \tag{2.53}$$

as indicated in Figure 2.3.

The components of $\boldsymbol{F}$ are given by

$$[\boldsymbol{F}] = \begin{bmatrix} 1 & K & 0 \\ 0 & 1 & 0 \\ 0 & 0 & 1 \end{bmatrix}. \tag{2.54}$$

This is a problem of plane motion where all the particles move parallel to the $0x_1x_2$ plane and the motion does not depend on x_3. Consequently, the rotation tensor $\boldsymbol{R}$ is of the form

$$[\boldsymbol{R}] = \begin{bmatrix} \cos\theta & \sin\theta & 0 \\ -\sin\theta & \cos\theta & 0 \\ 0 & 0 & 1 \end{bmatrix}, \tag{2.55}$$

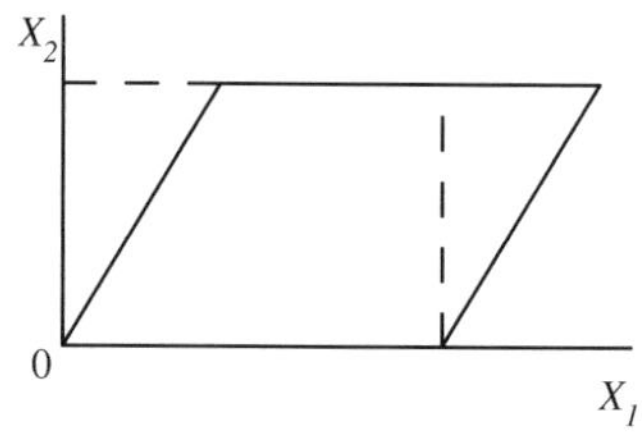

Figure 2.3. Simple shear deformation.

where $\boldsymbol{R}$ produces a clockwise rotation, when θ is measured in the clockwise direction and $\boldsymbol{R}$ operates on a vector in the X_1X_2plane.

We consider the form $(2.50)_2$ so that

$$\boldsymbol{V} = \boldsymbol{F}\boldsymbol{R}^{\mathrm{T}}$$

and it then follows from equations (2.54) and (2.55) that

$$\begin{bmatrix} V_{11} & V_{12} & V_{13} \\ V_{21} & V_{22} & V_{23} \\ V_{31} & V_{32} & V_{33} \end{bmatrix} = \begin{bmatrix} 1 & K & 0 \\ 0 & 1 & 0 \\ 0 & 0 & 1 \end{bmatrix} \begin{bmatrix} \cos\theta & -\sin\theta & 0 \\ \sin\theta & \cos\theta & 0 \\ 0 & 0 & 1 \end{bmatrix}$$
$$= \begin{bmatrix} \cos\theta + K\sin\theta & -\sin\theta + K\cos\theta & 0 \\ \sin\theta & \cos\theta & 0 \\ 0 & 0 & 1 \end{bmatrix}.$$

Since $V_{21} = V_{12}$,

$$-\sin\theta + K\cos\theta = \sin\theta.$$

Consequently,

$$\tan\theta = \frac{K}{2}.$$

The components of $\boldsymbol{V}$ and $\boldsymbol{R}$ are then given by

$$\begin{bmatrix} V_{11} & V_{12} \\ V_{21} & V_{22} \end{bmatrix} = \left(4 + K^2\right)^{-1/2} \begin{bmatrix} 2 + K^2 & -K + 2K \\ K & 2 \end{bmatrix}$$

and

$$V_{33} = 1,\ V_{31} = V_{32} = 0,$$
$$\begin{bmatrix} R_{11} & R_{12} \\ R_{21} & R_{22} \end{bmatrix} = \left(4 + K^2\right)^{-1/2} \begin{bmatrix} 2 & K \\ -K & 2 \end{bmatrix}.$$
$$R_{33} = 1,\ \mathrm{R}_{31} = R_{32} = 0.$$

Components of $\boldsymbol{U}$ can be obtained similarly.

2.10 Further Decomposition of $\boldsymbol{F}$

The deformation gradient $\boldsymbol{F}$ can be decomposed into the product of a dilatational part: and an isochoric or distortional part, as follows

$$\boldsymbol{F} = \left(J^{1/3}\boldsymbol{I}\right)\boldsymbol{F}^*. \tag{2.56}$$

Since $\det[\boldsymbol{F}^*] = 1$, $\boldsymbol{F}^*$ is the isochoric part and is known as the modified deformation gradient.

Other related decompositions that follow from equation (2.56) are

$$\boldsymbol{U} = \left(J^{1/3}\boldsymbol{I}\right)\boldsymbol{U}^* \text{ and } \boldsymbol{C} = \left(J^{2/3}\boldsymbol{I}\right)\boldsymbol{C}^*. \tag{2.57}$$

It may be deduced from equation $(2.57)_1$ that the stretches λ_α, $\alpha \in \{1, 2, 3\}$, that are the principal components of $\boldsymbol{U}$ can be similarly decomposed as

$$\lambda_\alpha = J^{1/3}\lambda^*_\alpha,$$

where $\lambda^*_1\lambda^*_2\lambda^*_3 = 1$ and the λ^*_α are known as the modified stretches.

2.11 Strain

The right and left stretch tensors $\boldsymbol{U}$ and $\boldsymbol{V}$, respectively, which result from the polar decomposition of the deformation gradient tensor $\boldsymbol{F}$, may be expressed in the form

$$\boldsymbol{U} = \left(\boldsymbol{F}^{\mathrm{T}}\boldsymbol{F}\right)^{1/2}, \ \boldsymbol{V} = \left(\boldsymbol{F}\boldsymbol{F}^{\mathrm{T}}\right)^{1/2}, \tag{2.58}$$

so that the squares of $\boldsymbol{U}$ and $\boldsymbol{V}$ are given by the positive definite symmetric tensors

$$\boldsymbol{C} = \boldsymbol{U}^2 = \boldsymbol{F}^{\mathrm{T}}\boldsymbol{F} \text{ and } \boldsymbol{B} = \boldsymbol{V}^2 = \boldsymbol{F}\boldsymbol{F}^{\mathrm{T}}, \tag{2.59}$$

or, in suffix notation,

$$C_{KL} = \frac{\partial x_i}{\partial X_K}\frac{\partial x_i}{\partial X_L} \quad \text{and} \quad B_{ij} = \frac{\partial x_i}{\partial X_K}\frac{\partial x_j}{\partial X_K}.$$

Tensors $\boldsymbol{C}$ and $\boldsymbol{B}$ are known as the right and left Cauchy strain tensors, respectively, and are a measure of the deformation at a material point in a body since the square of the length of a material line element in the spatial configuration is given by

$$dx_i dx_i = C_{KL} dX_K dX_L = d\boldsymbol{X} \cdot \boldsymbol{C} d\boldsymbol{X} \tag{2.60}$$

and in the reference configuration by

$$dX_K dX_K = B_{ij}^{-1} dx_i dx_j = d\boldsymbol{x} \cdot \boldsymbol{B}^{-1} d\boldsymbol{x}. \tag{2.61}$$

It follows that if $\boldsymbol{C} = \boldsymbol{I}$, which implies that $\boldsymbol{B} = \boldsymbol{I}$, for every material point in the body, the body has undergone a rigid body displacement.

Another measure of deformation is given by the tensor $\boldsymbol{E}$ known as Green's strain tensor and defined by

$$d\boldsymbol{x} \cdot d\boldsymbol{x} - d\boldsymbol{X} \cdot d\boldsymbol{X} = d\boldsymbol{X} \cdot \boldsymbol{E} d\boldsymbol{X} \tag{2.62}$$

$$dx_i dx_i - dX_L dX_L = 2E_{MN} dX_M dX_N.$$

It follows from equations (2.59), (2.60), and (2.62) that

$$\boldsymbol{E} = \frac{1}{2}(\boldsymbol{C} - \boldsymbol{I}) = \frac{1}{2}\left(\boldsymbol{F}^T\boldsymbol{F} - \boldsymbol{I}\right) = \frac{1}{2}\left(\boldsymbol{U}^2 - \boldsymbol{I}\right), \tag{2.63}$$

and, in suffix notation,

$$E_{KL} = \frac{1}{2}\left(\frac{\partial x_i}{\partial X_K}\frac{\partial x_i}{\partial X_L} - \delta_{KL}\right).$$

For a rigid rotation $d\boldsymbol{x} \cdot d\boldsymbol{x} = d\boldsymbol{X} \cdot d\boldsymbol{X}$, consequently $\boldsymbol{C} = \boldsymbol{I}$ and $\boldsymbol{E} = \boldsymbol{0}$.

The displacement vector $\boldsymbol{\xi}$ of a particle is given by

$$\boldsymbol{\xi} = \boldsymbol{x} - \boldsymbol{X}. \tag{2.64}$$

In suffix notation this equation presents a problem if we use the convention of an uppercase suffix referring to the reference configuration and a lower case suffix for the spatial configuration. This can be resolved as follows by expressing equation (2.64) in suffix notation as

$$\xi_i = x_i - X_K\delta_{Ki}, \tag{2.65}$$

where

$$\mathbf{E}_K \cdot \mathbf{e}_i = \delta_{Ki}.$$

It follows from equations $(2.63)_2$ and (2.65) that $\boldsymbol{E}$ may be expressed in the component form

$$E_{KL} = \frac{1}{2}\left[\frac{\partial\hat{\xi}_K}{\partial X_L} + \frac{\partial\hat{\xi}_L}{\partial X_K} + \frac{\partial\hat{\xi}_J}{\partial X_K}\frac{\partial\hat{\xi}_J}{\partial X_L}\right], \tag{2.66}$$

where $\hat{\xi}_K = \delta_{Ki}\xi_i$, and in this form $\boldsymbol{E}$ is sometimes called the Lagrangian strain.

A further measure of deformation is given by the tensor $\boldsymbol{\eta}$ defined by

$$dx_i dx_i - dX_K dX_K = 2\eta_{lm} dx_l dx_m = 2d\boldsymbol{x} \cdot \boldsymbol{\eta} d\boldsymbol{x}, \tag{2.67}$$

and it follows from equations (2.61) and (2.67) that

$$\boldsymbol{\eta} = \frac{1}{2}\left(\boldsymbol{I} - \boldsymbol{B}^{-1}\right) = \frac{1}{2}\left(\boldsymbol{I} - \boldsymbol{F}^{T^{-1}}\boldsymbol{F}\right), \tag{2.68}$$

or, in suffix notation,

$$\eta_{ij} = \frac{1}{2}\left[\delta_{ij} - \frac{\partial X_K}{\partial x_i}\frac{\partial X_K}{\partial x_j}\right],$$

and with the use of equation (2.65) this can be put in the form

$$\eta_{ij} = \frac{1}{2}\left[\frac{\partial \xi_i}{\partial x_j} + \frac{\partial \xi_j}{\partial x_i} - \frac{\partial \xi_s}{\partial x_i}\frac{\partial \xi_s}{\partial x_j}\right].$$

The tensor $\boldsymbol{\eta}$ is known as Almansi's strain tensor or the Eulerian strain tensor. It is less useful than $\boldsymbol{E}$. It is easily shown that

$$\boldsymbol{\eta} = \boldsymbol{F}^{T^{-1}}\boldsymbol{E}\boldsymbol{F}^{-1}.$$

It is useful to note that the components of the tensors $\boldsymbol{C}$ and $\boldsymbol{E}$ are independent of any rigid body rotation of the spatial configuration and the components of $\boldsymbol{B}$ and $\boldsymbol{\eta}$ are independent of any rotation of the reference configuration.

The material derivative of $\boldsymbol{E}$, Green's strain tensor, is related to $\boldsymbol{D}$, the rate of deformation tensor by

$$\dot{\boldsymbol{E}} = \boldsymbol{F}^T\boldsymbol{D}\boldsymbol{F}, \tag{2.69}$$

or, in suffix notation,

$$\dot{E}_{KL} = d_{ij}\frac{\partial x_i}{\partial X_K}\frac{\partial x_j}{\partial X_L}.$$

This result is obtained from equations (2.34) and (2.63) as follows, using symbolic notation,

$$\begin{aligned}\dot{\boldsymbol{E}} &= \frac{1}{2}\left(\frac{d}{dt}\left(\boldsymbol{F}^T\boldsymbol{F}-\mathbf{I}\right)\right) = \frac{1}{2}\left(\dot{\boldsymbol{F}}^T\mathbf{F}+\mathbf{F}^T\dot{\boldsymbol{F}}\right)\\ &= \frac{1}{2}\left(\boldsymbol{F}^T\boldsymbol{L}^T\boldsymbol{F}+\boldsymbol{F}^T\boldsymbol{L}\boldsymbol{F}\right) = \boldsymbol{F}^T\boldsymbol{D}\boldsymbol{F}.\end{aligned}$$

The tensor $\dot{\boldsymbol{E}}$ is known as the strain rate tensor, although sometimes $\boldsymbol{D}$ is called the strain rate tensor especially in plasticity theory. It is shown in the next section that $\boldsymbol{D} \simeq \dot{\boldsymbol{E}}$ when $|\partial u_i/\partial X_K| \ll 1$.

There are several other strain tensors in addition to the Green and Almansi strain tensors, for example, the Biot strain tensor,

$$\boldsymbol{E}_b = \boldsymbol{U} - \boldsymbol{I},$$

and the logarithmic strain tensors $\ln\mathbf{U}$ and $\ln\mathbf{V}$ whose principal components are $\ln\lambda_{\mathrm{i}}$.

2.12 Infinitesimal Strain

A tensor known as the infinitesimal strain tensor is a linearization of equation (2.66), given by

$$\boldsymbol{e} = \frac{1}{2}\left\{\nabla_{\mathrm{X}}\hat{\boldsymbol{\xi}} + \left(\nabla_{\mathrm{X}}\hat{\boldsymbol{\xi}}\right)^{\mathrm{T}}\right\} \tag{2.70}$$

where

$$\nabla_X\hat{\boldsymbol{\xi}} = \frac{\partial\hat{\xi}_K}{\partial X_L}\boldsymbol{E}_{\boldsymbol{K}} \otimes \boldsymbol{E}_{\boldsymbol{L}}$$

is the displacement gradient tensor.

For the present we will not express $\boldsymbol{e}$ in component form. The infinitesimal strain is related to Green's strain tensor by the relation obtained from equations (2.66) and (2.70)

$$\boldsymbol{E} = \boldsymbol{e} + \frac{1}{2}\left(\nabla_X\hat{\boldsymbol{\xi}}\right)^T\nabla_X\hat{\boldsymbol{\xi}}.$$

Consequently, if $\left|\nabla_X \hat{\boldsymbol{\xi}}\right|$ is, in some sense, small, it is reasonable to make the approximation $\boldsymbol{E} \simeq \boldsymbol{e}$ since

$$\boldsymbol{E} = \boldsymbol{e} + \mathrm{O}\left(\varepsilon^2\right) \tag{2.71}$$

where

$$\varepsilon = \left|\nabla_X \hat{\xi}\right|,$$

so that as $\varepsilon \to 0$, $\boldsymbol{E} \to \boldsymbol{e}$. It follows that if $\boldsymbol{\varepsilon} \ll 1$, $\boldsymbol{e}$ is a good approximation for $\boldsymbol{E}$. As $\varepsilon \to 0$, the difference between the spatial and reference configurations becomes negligible. Consequently, a linearization of equation (2.64a) is also a suitable infinitesimal strain tensor.

The tensor $\boldsymbol{e}$ is not a suitable measure for finite strain since $\boldsymbol{e}$ does not, in general, vanish for a rigid body motion. This may be shown as follows, suppose we have an infinitesimal rigid body rotation given by

$$\boldsymbol{F} = \boldsymbol{Q} = \boldsymbol{I} + \boldsymbol{\alpha}, |\boldsymbol{\alpha}| \ll 1$$

where $\boldsymbol{\alpha} = \nabla_X \hat{\boldsymbol{\xi}}$. The orthogonality condition $\boldsymbol{Q}\boldsymbol{Q}^{\mathrm{T}} = \boldsymbol{I}$ gives

$$\boldsymbol{\alpha} = -\boldsymbol{\alpha}^{\mathrm{T}} + \mathrm{O}\left(\varepsilon^2\right),$$

where $\varepsilon = |\boldsymbol{\alpha}|$. Green's strain tensor vanishes for a rigid displacement of the spatial configuration so that from equation (2.67) we have, for a rigid body rotation,

$$\boldsymbol{e} = \mathrm{O}\left(\varepsilon^2\right).$$

That is $\boldsymbol{e}$ vanishes to within an error $\mathrm{O}\left(\varepsilon^2\right)$ when $\boldsymbol{F} = \boldsymbol{Q}$ so that unless the rotation is infinitesimal $\boldsymbol{e}$ is not a suitable approximation for $\boldsymbol{E}$.

We now introduce the polar decomposition $\boldsymbol{F} = \boldsymbol{R}\boldsymbol{U}$ and put

$$\boldsymbol{R} = (\boldsymbol{I} + \boldsymbol{\alpha}) \quad \text{and} \quad \boldsymbol{U} = (\boldsymbol{I} + \boldsymbol{\beta})$$

noting that $\boldsymbol{\alpha}$ and $\boldsymbol{\beta}$ are both of $\mathrm{O}(\varepsilon)$. It follows from the orthogonality of $\boldsymbol{R}$ and the symmetry of $\boldsymbol{U}$ that

$$\boldsymbol{\alpha} = -\boldsymbol{\alpha}^{\mathrm{T}} + \mathrm{O}\left(\varepsilon^2\right), \boldsymbol{\beta} = \boldsymbol{\beta}^{\mathrm{T}}.$$

Then we have

$$\nabla_X \hat{\boldsymbol{\xi}} = \boldsymbol{F} - \boldsymbol{I} = \boldsymbol{\alpha} + \boldsymbol{\beta} + \mathrm{O}\left(\varepsilon^2\right),$$

that is, for infinitesimal deformation, when terms $\mathrm{O}(\varepsilon^2)$ can be neglected, the tensor $\boldsymbol{\beta}$ is the symmetric part of the displacement gradient tensor and is identical to $\boldsymbol{e}$ given by equation (2.61). The antisymmetric tensor $\boldsymbol{\alpha}$ can be regarded as a measure of the infinitesimal rotation.

It should be noted that the use of infinitesimal strain tensor $\boldsymbol{e}$ is justified when $|\nabla_X \boldsymbol{U}| = 1$, and this implies that both the rotation and $\boldsymbol{E}$ are small not just the $\boldsymbol{E}$. When the use of the infinitesimal tensor is justified when terms $\mathrm{O}(\varepsilon^2)$ can be neglected, the distinction between spatial and material coordinates can be neglected and we can express $\boldsymbol{e}$ and $\boldsymbol{\alpha}$ in coordinate form as

$$e_{ij} = \frac{1}{2}\left(\frac{\partial u_i}{\partial x_j} + \frac{\partial u_j}{\partial x_i}\right), \alpha_{ij} = \frac{1}{2}\left(\frac{\partial u_i}{\partial x_j} - \frac{\partial u_j}{\partial x_i}\right). \tag{2.72}$$

The theory for the infinitesimal strain and rotation tensors is analogous to the theory for the rate of deformation tensor $\boldsymbol{D}$ and the spin tensor $\boldsymbol{W}$ with the infinitesimal displacement analogous to the velocity.

When the use of infinitesimal strain approximation (2.69) is justified it is evident that

$$\dot{\mathrm{e}}_{ij} \approx \mathrm{d}_{ij}.$$

This is perhaps the reason why $\boldsymbol{D}$ is often called the strain rate tensor.

EXERCISES

2.1. If $\boldsymbol{da}$ and $\boldsymbol{dA}$ are the vectors representing an element of area in the spatial and reference configurations, respectively, show that

$$da_i = \frac{\rho_0}{\rho} \frac{\partial X_K}{\partial x_i} dA_K,$$

or, in symbolic notation,

$$\boldsymbol{da} = \frac{\rho_0}{\rho} \boldsymbol{F}^{-1^T} \boldsymbol{dA}.$$

This relation is known as Nanson's formula

(a) Use Nanson's formula to show that $\mathrm{Div}(\boldsymbol{JF}^{-1}) = 0$.

Note that, for a closed surface enclosing volume $V = \int_{\partial V} \mathbf{da} = \mathbf{0}$.

2.2. Prove the following results

(a)

$$\frac{d}{dt}\left(\frac{\partial X_K}{\partial x_i}\right) = -\frac{\partial v_j}{\partial x_i}\frac{\partial X_K}{\partial x_j}.$$

(b)

$$\frac{\partial}{\partial x_i}\left(J^{-1}F_{iK}\right) = 0.$$

(c)

$$\frac{\partial U_K}{\partial X_K} = J\frac{\partial}{\partial x_i}\left(J^{-1}F_{iK}U_K\right).$$

2.3. Let $\boldsymbol{dx}^{(1)}$ and $\boldsymbol{dx}^{(2)}$ be material line elements emanating from a particle P. They specify a material surface element given by

$$\boldsymbol{da} = \boldsymbol{dx}^{(1)} \times \boldsymbol{dx}^{(2)}.$$

Show that

$$\frac{d}{dt}(da_i) = \frac{\partial v_j}{\partial x_j} da_i - \frac{\partial v_j}{\partial x_i} da_j.$$

2.4. If $\boldsymbol{F} = \boldsymbol{RU}$ is the polar decomposition of the deformation gradient show that

$$\mathbf{W} = \dot{\boldsymbol{R}}\boldsymbol{R}^{\mathrm{T}} + \frac{1}{2}\left\{\boldsymbol{R}\dot{\boldsymbol{U}}\boldsymbol{U}^{-1}\boldsymbol{R}^{\mathrm{T}} - \left(\boldsymbol{R}\dot{\boldsymbol{U}}\boldsymbol{U}^{-1}\boldsymbol{R}^{\mathrm{T}}\right)^{\mathrm{T}}\right\}$$

and

$$\boldsymbol{D} = \frac{1}{2}\boldsymbol{R}\left\{\dot{\boldsymbol{U}}\boldsymbol{U}^{-1} + \boldsymbol{U}^{-1}\dot{\boldsymbol{U}}\right\}\boldsymbol{R}^{T}.$$

2.5. If $\boldsymbol{e}$ is the isotropic part of the infinitesimal strain tensor show that

$$J = 1 + 3e + 0\left(\varepsilon^2\right),$$

where $\varepsilon = |\nabla_X \boldsymbol{U}|$.

2.6. Show that the acceleration vector can be expressed in the form

$$\boldsymbol{a} = \frac{\partial \boldsymbol{v}}{\partial t} + (\nabla \otimes \boldsymbol{v})\boldsymbol{v} - (\nabla \otimes v)^{\mathrm{T}}\boldsymbol{v} + \frac{1}{2}\nabla \boldsymbol{v} \cdot \boldsymbol{v}.$$

2.7. Show that

$$\rho\left(\frac{\partial \boldsymbol{v}}{\partial t} + \boldsymbol{v} \cdot \mathrm{grad}\boldsymbol{v}\right) = \frac{\partial(\rho \boldsymbol{v})}{\partial t} + \mathrm{div}(\rho \boldsymbol{v} \otimes \boldsymbol{v}).$$

2.8. Prove that, for each particle of a deforming continuum there is at least one material line element whose direction is instantaneously stationary.

2.9. Obtain the polar decompositions $\boldsymbol{F} = \boldsymbol{R}\boldsymbol{U}$ and $\boldsymbol{F} = \boldsymbol{V}\boldsymbol{R}$ for the deformation

$$[\boldsymbol{F}] = \begin{bmatrix} 1 & \frac{3}{2} & 0 \\ 0 & 1 & 0 \\ 0 & 0 & 1 \end{bmatrix}.$$

3 Stress

3.1 Contact Forces and Body Forces

In order to introduce the concept of stress we first consider the spatial configuration of a body that occupies a region v of Euclidean point space E_3. At an interior point P($\boldsymbol{x}$), of the body, there is an infinite number of plane elements that contain P. Let $\boldsymbol{n}$ be the unit normal to one of these elements with area Δa as indicated in Figure 3.1.

The infinitesimal force $\Delta\boldsymbol{F}$ acts on a surface element $\Delta a\boldsymbol{n} = \Delta\mathbf{a}$. Denote by m_+ the material adjacent to the element on the side from which $\boldsymbol{n}$ is directed and m_- the material on the other side. In general m_+ exerts a force $\Delta\boldsymbol{F}$ on m_-, and m_- exerts an equal in magnitude but opposite in direction force on m_+. The stress vector at P acting on the plane element with unit vector $\boldsymbol{n}$ is defined as

$$\boldsymbol{t}(\boldsymbol{n},\boldsymbol{x},t) = \lim_{\Delta a\to 0}\frac{\Delta\boldsymbol{F}}{\Delta a}.$$

For convenience we suppress the dependence on $\boldsymbol{x}$ and time t. The stress vector $\boldsymbol{t}$ is known as a contact force. At present we assume that $\boldsymbol{t}$ depends continuously on $\boldsymbol{n}$ so that

$$\boldsymbol{t}(\boldsymbol{n}) = -\boldsymbol{t}(-\boldsymbol{n}).$$

This condition is not satisfied when we have singular surfaces across which discontinuities of certain field variables can occur, for example, for shock waves.

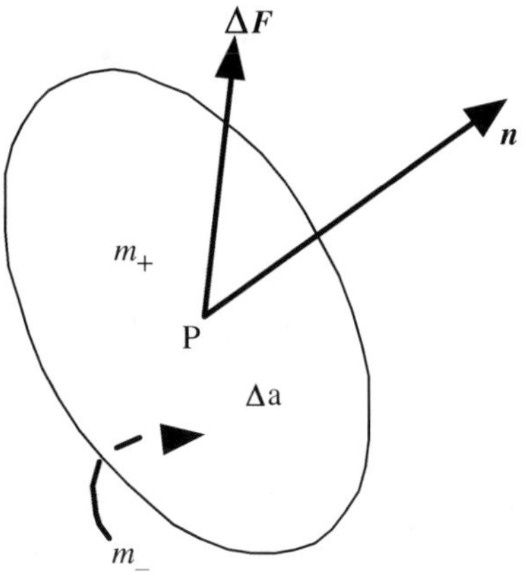

Figure 3.1. Force on area element at point $P(x)$.

The resultant moment exerted by m_+ on m_- is denoted by $\Delta \boldsymbol{M}$ and we assume that at P,

$$\lim_{\Delta a \to 0} \frac{\Delta \boldsymbol{M}}{\Delta \mathrm{a}} = 0,$$

so that contact forces are present but not contact torques. In certain media, known as polar media, contact torques can exist; however, in this book we will not be concerned with this aspect of continuum mechanics.

The stress vector at points on the surface $\partial \mathrm{v}$ of a body, with $\boldsymbol{n}$ the outward unit normal to the surface, is known as the surface traction.

The stress vector can be resolved into a normal component

$$N(\boldsymbol{n}) = \boldsymbol{n} \cdot \mathbf{t}(\boldsymbol{n}), \tag{3.1}$$

and a shearing component

$$S(\boldsymbol{n}) = \left\{ |\boldsymbol{t}(\boldsymbol{n})|^2 - N(\boldsymbol{n})^2 \right\}^{1/2}. \tag{3.2}$$

When N is positive the normal component is said to be tensile and when N is negative it is said to be compressive, although the opposite convention is often used in soil mechanics.

Forces acting on interior particles, due to external agencies such as gravity, are known as body forces, and we denote the body force per unit mass by $\boldsymbol{f}$. It is assumed that there are no body torques.

Henceforth, some equations are given in both symbolic and Cartesian tensor notation.

In what follows the density, denoted by ρ, is a scalar property with physical dimensions of mass per unit volume. As usual the mass of a body is a measure of the amount matter of the body.

According to a generalization of Newton's second law the resultant force acting on a body, due to surface tractions and body forces, is equal to the rate of change of momentum of the body. Consequently,

$$
\begin{aligned}
\int_{\partial \mathrm{v}} \boldsymbol{t} da + \int_{\mathrm{v}} \rho \boldsymbol{f} d\mathrm{v} &= \frac{d}{dt} \int_{\mathrm{v}} \rho \boldsymbol{v} d\mathrm{v}, \\
\int_{\partial \mathrm{v}} t_i da + \int_{\mathrm{v}} \rho f_i d\mathrm{v} &= \frac{d}{dt} \int_{\mathrm{v}} \rho \mathrm{v}_i d\mathrm{v},
\end{aligned}
\tag{3.3}
$$

where $\boldsymbol{v}(\boldsymbol{x}, t)$ is the velocity, $\rho(\boldsymbol{x}, t)$ is the density, $d\mathrm{v}$ is an element of volume, and da is an element of surface area. It is shown in chapter 2 that

$$
\frac{d}{dt} \int_{\mathrm{v}} \rho \boldsymbol{v} d\mathrm{v} = \int_{\mathrm{v}} \rho \frac{d\boldsymbol{v}}{dt} d\mathrm{v}.
$$

The resultant moment acting on a body, about a fixed point in an inertial frame of reference or about the mass center of the body, due to surface tractions and body forces, is equal to the rate of change of momentum of the body. Consequently,

$$
\begin{aligned}
\int_{\partial \mathrm{v}} \boldsymbol{x} \times \boldsymbol{t} da + \int_{\mathrm{v}} \boldsymbol{x} \times \rho \boldsymbol{f} d\mathrm{v} &= \frac{d}{dt} \int_{\mathrm{v}} \boldsymbol{x} \times \rho \boldsymbol{v} d\mathrm{v}, \\
\int_{\partial \mathrm{v}} e_{ijk} x_j t_k da + \int_{\mathrm{v}} e_{ijk} x_j \rho f_k d\mathrm{v} &= \frac{d}{dt} \int_{\mathrm{v}} e_{ijk} x_j \rho v_k d\mathrm{v},
\end{aligned}
\tag{3.4}
$$

where $\boldsymbol{x}$ is the position vector with respect to the fixed point or the mass center.

3.2 Cauchy or True Stress

Consider a plane element at P whose normal is in the x_1 direction. Then,

$$\boldsymbol{t}(\mathbf{e}_1) = \sigma_{11}\mathbf{e}_1 + \sigma_{12}\mathbf{e}_2 + \sigma_{13}\mathbf{e}_3 = \sigma_{1i}\mathbf{e}_{1i}, \tag{3.5}$$

where σ_{11}, σ_{12}, and σ_{13} are the x_1, x_2, and x_3 components, respectively, of the stress vector acting on the plane element; σ_{11} is the normal component, and σ_{12} and σ_{13} are the shear components. It follows that the resulting shearing stress on the element is

$$S(\mathbf{e}_1) = \left(\sigma_{12}^2 + \sigma_{13}^2\right)^{1/2}.$$

Similarly for plane elements at P whose normals are in the x_2 and x_3 directions,

$$\boldsymbol{t}(\mathbf{e}_2) = \sigma_{21}\mathbf{e}_1 + \sigma_{22}\mathbf{e}_2 + \sigma_{23}\mathbf{e}_3, \tag{3.6}$$

$$\boldsymbol{t}(\mathbf{e}_3) = \sigma_{31}\mathbf{e}_1 + \sigma_{32}\mathbf{e}_2 + \sigma_{33}\mathbf{e}_3, \tag{3.7}$$

$$S(\mathbf{e}_2) = \left(\sigma_{21}^2 + \sigma_{23}^2\right)^{1/2},$$

$$S(\mathbf{e}_3) = \left(\sigma_{32}^2 + \sigma_{31}^2\right)^{1/2}.$$

Equations (3.5) to (3.7) may be expressed in suffix notation as

$$\boldsymbol{t}(\mathbf{e}_i) = \sigma_{ij}\mathbf{e}_j. \tag{3.8}$$

The matrix $\left[\sigma_{ij}\right]$ specifies the state of stress at P. Component σ_{ij} is defined as the stress component in the x_j direction of the stress vector acting on the plane element whose normal is in the x_i direction. Some texts define this as σ_{ji} rather than σ_{ij}. The signs of the components σ_{ij} are obtained as follows. If the normal to a plane element is in the $\pm\, x_i$ direction and the stress vector component is in the $\pm\, x_j$ directions, the component σ_{ij} is positive, otherwise it is negative. Positive stress components are shown in Figure 3.2 for a plane state of stress. Now consider an arbitrary area element with unit normal $\boldsymbol{n}$ and stress vector $\boldsymbol{t}(\boldsymbol{n})$ acting on it. Let this element be the face ABC of the infinitesimal tetrahedron PABC with sides PA, PB, and PC in the $x_1, x_2,$ and x_3 directions, respectively, as shown in Figure 3.3.

Also let Δa be the area of ABC and h be the perpendicular distance of P from ABC. The components of $\boldsymbol{t}$ are given by $t_i = \boldsymbol{t}(\boldsymbol{n})\,\mathbf{e}_i$. Then, the equation of motion of the tetrahedron in the x_i direction is

$$t_i \Delta \mathrm{a} + \rho\left(\frac{1}{3}h\Delta \mathrm{a}\right) f_i - (\sigma_{1i}\mathrm{PCB} + \sigma_{2i}\mathrm{PAC} + \sigma_{3i}\mathrm{PAB}) = \rho\left(\frac{1}{3}h\Delta \mathrm{a}\right)\dot{v}_i, \quad (3.9)$$

where $\dot{\boldsymbol{v}} = \dfrac{dv}{dt} = \dot{v}_i \mathbf{e}_i = \dfrac{dv_i}{dt}\mathbf{e}_i$ is the particle acceleration.

Substituting

$$\mathrm{PCB} = n_1 \Delta \mathrm{a},\ \mathrm{PAC} = n_2 \Delta \mathrm{a},\ \mathrm{PAB} = n_3 \Delta \mathrm{a},$$

in equation (3.9) and letting $h \to 0$ gives

$$t_i = \sigma_{ji} n_j, \quad \boldsymbol{t} = \sigma^T \boldsymbol{n}. \quad (3.10)$$

The elements of the matrix $[\sigma_{ji}]$ are the components of a second-order tensor $\boldsymbol{\sigma}^T$ since $\sigma_{ji} n_j$ are the components of a vector for arbitrary unit vector

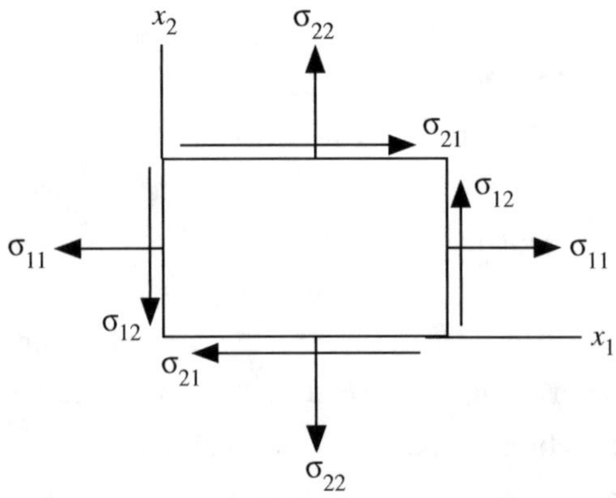

Figure 3.2. Stress components for plane Cauchy stress.

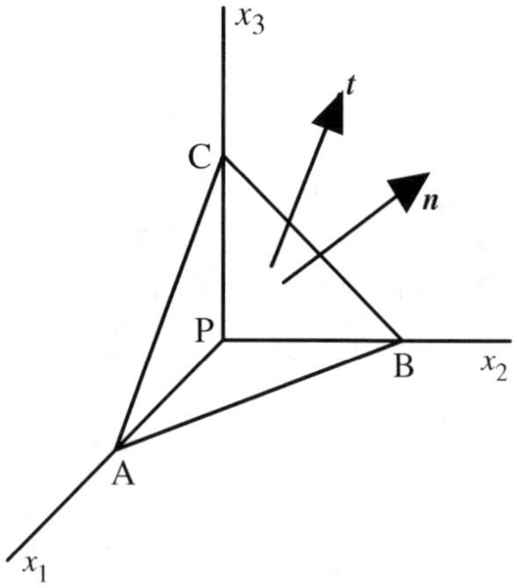

Figure 3.3. Equilibrium of infinitesimal. Tetrahedron PABC.

$\boldsymbol{n}$. This follows from the test for tensor character given in chapter 1 since $\boldsymbol{\sigma}^T$ is a tensor so is $\boldsymbol{\sigma}$. If σ_{ij} and σ'_{ij} are the components of stress referred to the rectangular Cartesian coordinate systems Ox_i and Ox'_i, respectively, it follows from the tensor transformation rule

$$\sigma'_{ij} = a_{ik}a_{jl}\sigma_{kl}, \tag{3.11}$$

$$\sigma_{kl} = a_{ik}a_{jl}\sigma'_{ij}, \tag{3.12}$$

where a_{ik} are the elements of the transformation matrix $[a]$. Equations (3.11) and (3.12) can be expressed in symbolic matrix notation as

$$[\boldsymbol{\sigma}'] = [\boldsymbol{a}][\boldsymbol{\sigma}][\boldsymbol{a}]^T,$$

$$[\boldsymbol{\sigma}] = [\boldsymbol{a}]^T[\boldsymbol{\sigma}'][\boldsymbol{a}],$$

respectively.

Equations (3.1) and (3.10) give the expression

$$N(\boldsymbol{n}) = \sigma_{ji}n_jn_1 = \boldsymbol{n}\cdot\boldsymbol{\sigma}^T\boldsymbol{n} \tag{3.13}$$

for the normal stress acting on a plane element with unit normal $\boldsymbol{n}$ and the resultant shearing stress is given by equation (3.2).

The stress tensor

$$\boldsymbol{\sigma} = \sigma_{ij}\mathbf{e}_i \otimes \mathbf{e}_j$$

is referred to the current or spatial configuration of the body and is known as the Cauchy stress and is a particular case of the class of Eulerian tensors, that is, the class of tensors referred to the spatial configuration. We have yet to show that it is symmetric for nonpolar media.

3.3 Spatial Form of Equations of Motion

Substituting equation (3.10) in equation (3.3) gives

$$\int_{\partial \mathrm{v}} \sigma_{ji} n_j \mathrm{da} + \int_{\mathrm{v}} \rho(f_i - \dot{v}_i) d\mathrm{v} = 0,$$

and application of the divergence theorem gives

$$\int_{\mathrm{v}} \left\{ \frac{\partial \sigma_{ji}}{\partial x_j} + \rho(f_i - \dot{v}_i) \right\} d\mathrm{v} = 0. \tag{3.14}$$

Since equation (3.14) must hold for an arbitrary volume and it is assumed that the integrand is continuous,

$$\begin{aligned} \frac{\partial \sigma_{ji}}{\partial x_j} + \rho f_i &= \rho \frac{dv_i}{dt}, \\ \mathrm{div}\boldsymbol{\sigma}^T + \rho \boldsymbol{f} &= \rho \frac{d\boldsymbol{v}}{dt}. \end{aligned} \tag{3.15}$$

It should be noted that $d\boldsymbol{v}/dt = \partial \boldsymbol{v}/\partial t + \boldsymbol{v}.\mathrm{grad}\boldsymbol{v}$, $\mathrm{d}v_i/\mathrm{d}t = \partial v_i/\partial t + v_j \partial v_i/\partial x_j$, is the material derivative of $\boldsymbol{v}$. For linearized problems $\mathrm{d}\boldsymbol{v}/\mathrm{d}t$ is replaced by $\partial \boldsymbol{v}/\partial t$.

Similarly the angular momentum equation with (3.10) substituted in equation (3.4) is

$$\int_{\partial \mathrm{v}} \mathrm{e}_{ijk} x_j \sigma_{rk} n_r da + \int_{\mathrm{v}} \mathrm{e}_{ijk} x_j \rho(f_k - \dot{v}_k) d\mathrm{v} = 0 \tag{3.16}$$

in Cartesian tensor suffix notation. With the use of the divergence theorem, the substitution property of the Kronecker delta and equation (3.15), equation (3.16) becomes

$$\int_{\mathrm{V}} \mathrm{e}_{ijk}\sigma_{jk}d\mathrm{v} = 0. \tag{3.17}$$

Since equation (3.17) must hold for an arbitrary volume and it is assumed that the integrand is continuous, $\mathrm{e}_{ijk}\sigma_{jk} = 0$. It then follows that

$$\sigma_{jk} = \sigma_{kj}, \quad \boldsymbol{\sigma} = \boldsymbol{\sigma}^{\boldsymbol{T}},$$

which shows that the stress tensor $\boldsymbol{\sigma}$ is symmetric in the absence of body moments.

3.4 Principal Stresses and Maximum Shearing Stress

We have shown that the Cauchy stress at a point is completely described by a second-order symmetric tensor $\boldsymbol{\sigma}$. The properties of symmetric second-order tensors have been given in detail in chapter 1, and some of these properties are repeated here in less detail for the Cauchy stress tensor.

If the direction of the stress vector acting on a plane element is normal to the element, there is no shearing stress acting on the element and the stress is a principal stress. It then follows that

$$\sigma_{ij}n_j = \lambda n_i, \quad \boldsymbol{\sigma n} = \lambda \boldsymbol{n}, \tag{3.18}$$

where λ is an eigenvalue and $\boldsymbol{n}$ is an eigenvector.

Since $\boldsymbol{\sigma}$ is symmetric it has three real eigenvalues, $\sigma_\alpha, \alpha \in \{1,2,3\}$, which may or may not be distinct. These eigenvalues of $\boldsymbol{\sigma}$ are known as principal stresses, and three corresponding eigenvectors are known as principal directions and are denoted by the orthonormal triad $\boldsymbol{n}^{(\alpha)}, \alpha \in \{1,2,3\}$. It should be noted that the superscript α involves no summation. Equation (3.18) can be put in the form

$$(\sigma_{ij} - \lambda\delta_{ij})n_j = 0, \quad (\boldsymbol{\sigma} - \lambda\boldsymbol{I})\boldsymbol{n} = \boldsymbol{0}, \tag{3.19}$$

and it follows that $\sigma^{(\alpha)}, \alpha \in \{1,2,3\}$ are the roots of the cubic equation

$$\det\left[\sigma_{ij} - \lambda\delta_{ij}\right] = 0. \tag{3.20a}$$

Equation (3.20) can be expanded as

$$\lambda^3 - I_1\lambda^2 + I_2\lambda - I_3 = 0,$$

where

$$I_1 = \sigma_{ii}, I_2 = \frac{1}{2}\left\{(\sigma_{ii})^2 - \sigma_{ij}\sigma_{ij}\right\}, I_3 = \det\left[\sigma_{ij}\right],$$
$$I_1 = \mathrm{tr}\boldsymbol{\sigma}, I_2 = \frac{1}{2}\left\{(\mathrm{tr}\boldsymbol{\sigma})^2 - \mathrm{tr}\boldsymbol{\sigma}^2\right\}, I_3 = \det[\boldsymbol{\sigma}]$$

are independent of the choice of coordinate system and are known as the first, second, and third basic invariants, respectively, of the stress tensor $\boldsymbol{\sigma}$. If the third basic invariant $I_3 = 0$, then the tensor $\boldsymbol{\sigma}$ is singular and one of the principal stresses is zero.

The components of $\boldsymbol{n}^{(\alpha)}, \alpha \in \{1,2,3\}$, corresponding to σ_α, can be obtained from any two of equations (3.19) and $\left|\boldsymbol{n}^{(\alpha)}\right| = 1$ when the σ_α have been obtained.

When the σ_α are distinct, the spectral representation of the Cauchy stress tensor is

$$\boldsymbol{\sigma} = \sum_{\alpha=1}^{3} \sigma_\alpha \boldsymbol{n}^{(\alpha)} \otimes \boldsymbol{n}^{\alpha}, \alpha \in \{1,2,3\},$$

where the summation convention is suspended. If $\sigma_1 = \sigma_2 \neq \sigma_3$ any direction normal to $\boldsymbol{n}^{(3)}$ is principal direction and the spectral representation is

$$\boldsymbol{\sigma} = \sigma_3\boldsymbol{n}^{(3)} \otimes \boldsymbol{n}^{(3)} + \sigma_1\left(\boldsymbol{I} - \boldsymbol{n}^{(3)} \otimes \boldsymbol{n}^{(3)}\right).$$

If $\sigma_1 = \sigma_2 = \sigma_3 = -p$, we have a hydrostatic or isotropic state of stress and the spectral representation is

$$\sigma = -p\boldsymbol{I}, \quad \sigma_{ij} = -p\delta_{ij}.$$

Equation (3.19) can also be obtained by considering the stationary values, with respect to $\boldsymbol{n}$, of the normal stress $N(\boldsymbol{n})$. In order to determine these directions we must find the stationary values of $\sigma_{ij}n_in_j$ subject to the subsidiary condition $n_in_i = 1$. By introducing the Lagrangian multiplier λ we have

$$\frac{\partial}{\partial n_k}\left(\sigma_{ij}n_in_j - \lambda n_in_i\right) = 0,$$

and differentiation gives

$$\left(\sigma_{ij} - \lambda\delta_{ij}\right)n_i = 0. \tag{3.20b}$$

that is identical to equation (3.19). The Lagrangian multiplier λ is identified with a principal stress and the normal stress $N(\boldsymbol{n})$ is stationary and is a principal stress, when $\boldsymbol{n}$ is an eigenvector of $\boldsymbol{\sigma}$. It follows that the greatest and least normal stresses at a point are principal stresses.

The directions associated with the stationary values of the magnitude of the shearing stress, $|S(\boldsymbol{n})|$, are determined in a similar manner except that in this case it is desirable to express equation (3.2) with respect to axes coinciding with the principal directions of $\boldsymbol{\sigma}$, that is,

$$S^2(\boldsymbol{n}) = (\sigma_1 n_1)^2 + (\sigma_2 n_2)^2 + (\sigma_3 n_3)^2 - \left(\sigma_1 n_1^2 + \sigma_2 n_2^2 + \sigma_3 n_3^2\right)^2. \tag{3.21}$$

It is evident from equation (3.21) that if $n_1 = \pm 1, n_2 = n_3 = 0$; $n_2 = \pm 1, n_1 = n_3 = 0$; or $n_3 = \pm 1, n_1 = n_2 = 0$, $|S(\boldsymbol{n})| = 0$, which is as expected since the shearing stresses acting on plane elements normal to the principal directions are zero. We introduce a Lagrangian multiplier λ to obtain the directions associated with other stationary values of $S^2(\boldsymbol{n})$ given by equation (3.21) and subject to the constraint $n_in_i = 1$. Then stationary values, in addition to the zero values on the plane elements whose normals are in the principal directions, are obtained from

$$\frac{\partial}{\partial n_k}\left(S^2 - \lambda n_in_i\right) = 0,$$

where S^2 is given by equation (3.21). After some computation we obtain the following components, referred to the principal axes of $\boldsymbol{\sigma}$, of the corresponding stationary values of $|S|$,

$$\begin{aligned} n_1 &= 0, n_2 = \sqrt{1/2}, n_3 = \sqrt{1/2}, |S| = \frac{1}{2}|\sigma_2 - \sigma_3|, \\ n_1 &= \sqrt{1/2},\ n_2 = 0, n_3 = \sqrt{1/2}, |S| = \frac{1}{2}|\sigma_3 - \sigma_1|, \\ n_1 &= \sqrt{1/2},\ n_2 = \sqrt{1/2}, n_3 = 0, |S| = \frac{1}{2}|\sigma_1 - \sigma_2|. \end{aligned} \tag{3.22}$$

It follows from equation (3.22) that the maximum value of $|S|$ is given by

$$|S|_{\max} = \frac{1}{2}(\sigma_{\max} - \sigma_{\min}), \tag{3.23}$$

where $\sigma_{\max}$ and $\sigma_{\min}$ are the algebraically greatest and least principal stresses, respectively. The result (3.23) has an important application in the theory of plasticity since the Tresca yield condition states that the onset of plastic deformation of a metal occurs when $|S|_{\max}$ reaches a certain critical value, which is equal to the yield point in simple shear.

3.5 Pure Shear Stress and Decomposition of the Stress Tensor

A state of pure shear stress may be defined as follows. If a rectangular Cartesian coordinate system exists such that the normal components of stress at a point are zero, that is, $\sigma_{11} = \sigma_{22} = \sigma_{33} = 0$, the state of stress at that point is a state of pure shear. The stress tensor corresponding to a state of pure shear may be expressed in the form

$$\begin{aligned} \boldsymbol{\sigma} &= \sigma_{12}(\mathbf{e}_1 \otimes \mathbf{e}_2 + \mathbf{e}_2 \otimes \mathbf{e}_1) + \sigma_{23}(\mathbf{e}_2 \otimes \mathbf{e}_3 + \mathbf{e}_3 \otimes \mathbf{e}_2) \\ &\quad + \sigma_{31}(\mathbf{e}_3 \otimes \mathbf{e}_1 + \mathbf{e}_1 \otimes \mathbf{e}_3), \end{aligned} \tag{3.24}$$

where the $\mathbf{e}_i$ are the unit base vectors of a coordinate system for which $\sigma_{11} = \sigma_{22} = \sigma_{33} = 0$. This coordinate system is not unique since the form (3.24) is valid for an infinite number of coordinate systems. The proof of this is given by Pearson [1]. It is clear that $\sigma_{ii} = \mathrm{tr}\boldsymbol{\sigma}$ is a necessary condition for a state of pure shear stress and the proof of this is also a sufficient condition is also given by Pearson [1].

We now introduce the decomposition of the stress tensor into its isotropic or hydrostatic part and deviatoric parts

$$\sigma_{ij} = s_{ij} + \frac{\sigma_{kk}}{3}\delta_{ij}, \quad \boldsymbol{\sigma} = \boldsymbol{s} + \frac{\mathrm{tr}\boldsymbol{\sigma}}{3}\boldsymbol{I}, \tag{3.25}$$

where $\frac{\mathrm{tr}\boldsymbol{\sigma}}{3}\boldsymbol{I}$ and $\boldsymbol{s}$ are the isotropic and deviatoric parts, respectively, of the stress tensor. It is easily verified that $s_{ii} = \mathbf{tr}\boldsymbol{s} = 0$ and that the basic invariants of the tensor $\boldsymbol{s}$ are

$$J_1 = 0, \quad J_2 = -\frac{1}{2}\left(s_{ij}s_{ij}\right), \quad J_3 = \det\left[s_{ij}\right],$$

Consequently, the deviatoric part of $\boldsymbol{\sigma}$ represents a state of pure shear. The isotropic part corresponds to a pure hydrostatic stress with the normal stress the same for all plane elements at a point and no shearing stress. The decomposition equation (3.25) is very important in plasticity theory since one interpretation of the Mises yield condition is that yielding of a metal occurs when $|J_2|$ reaches a critical value.

A special case of a deviatoric stress is given by

$$\boldsymbol{\sigma} = \tau\mathbf{e}_1 \otimes \mathbf{e}_2,$$

as indicated in Figure 3.4.

This stress state is often described as pure shear; however, we prefer to regard it as a special case of pure shear and call it simple shear.

3.6 Octahedral Shearing Stress

First we consider a rectangular Cartesian coordinate system with the origin at a point P in a material body and with its axes coinciding with the principal axes of the stress tensor $\boldsymbol{\sigma}$ at P. A plane, which makes equal intercepts

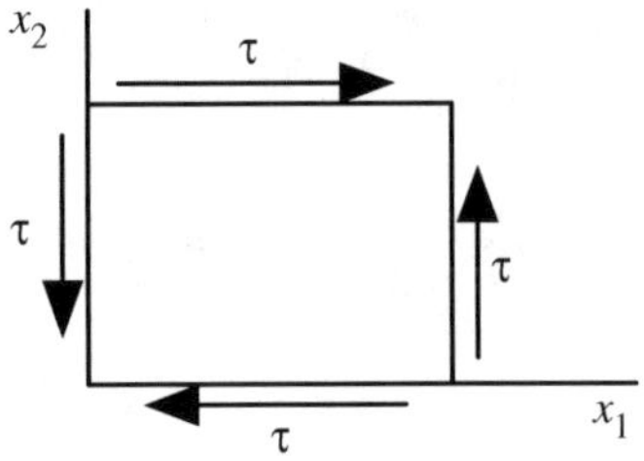

Figure 3.4. Simple shear stress.

with these three coordinate axes, is known as an octahedral plane. An octahedral plane, which intercepts the positive directions of the axes, is shown in Figure 3.5.

The normals to the eight octahedral planes are given by

$$\boldsymbol{n} = \frac{1}{\sqrt{3}}(\pm\mathbf{e}_1 \pm\mathbf{e}_2 \pm\mathbf{e}_3). \tag{3.26}$$

Now imagine an octahedron enclosed by the eight elemental octahedral planes and shrunk to point P and consider eight plane elements whose normals are given by equation (3.26). Since the axes coincide with the principal axes of stress, it follows from equations (3.13) and (3.26) that the same normal stress

$$N_{oct} = \frac{1}{3}(\sigma_1 + \sigma_2 + \sigma_3) = \frac{I_1}{3},$$

that is, the isotropic part of the stress tensor, acts on each octahedral plane. The shearing stress $|S_{oct}|$ acting on a plane element with a normal given by equation (3.26) is known as an octahedral shearing stress, and it follows from equations (3.21) and (3.26) that it is the same for all eight octahedral planes and is given by

$$S_{oct}^2 = \frac{1}{3}\left(\sigma_1^2 + \sigma_2^2 + \sigma_3^2\right) - \frac{1}{9}(\sigma_1 + \sigma_2 + \sigma_3)^2, \tag{3.27}$$

which can be put in the form

$$S_{oct}^2 = \frac{1}{9}\left\{(\sigma_1 - \sigma_2)^2 + (\sigma_2 - \sigma_3)^2 + (\sigma_3 - \sigma_1)^2\right\}. \tag{3.28}$$

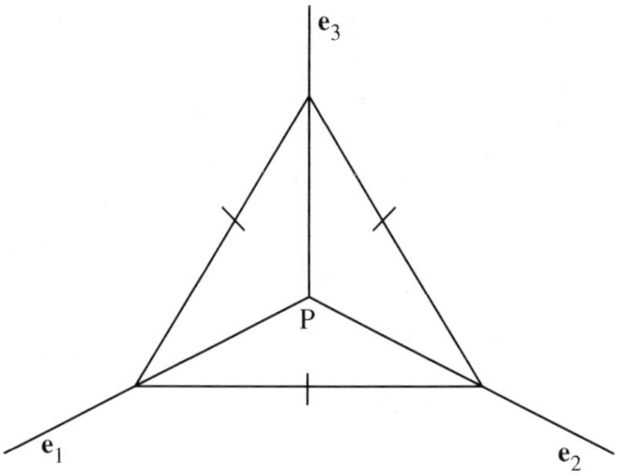

Figure 3.5. Octahedral plane.

In equation (3.28) the principal components of $\boldsymbol{\sigma}$ can be replaced by the principal components of $\boldsymbol{s}$ so that equation (3.27) becomes

$$S_{oct}^2 = \frac{1}{3}\left(s_1^2 + s_2^2 + s_3^2\right),$$

which referred to an arbitrary coordinate system gives

$$S_{oct}^2 = \frac{1}{3} s_{ij} s_{ij} = -\frac{2}{3} J_2. \tag{3.29}$$

Equation (3.29) has an important application in the theory of plasticity [2]. The von Mises yield criterion implies that yielding of a metal is not influenced by the isotropic part of the stress tensor, and yielding occurs when $|J_2|$ reaches a critical value. This criterion can be expressed in the alternative forms [2],

$$s_{ij} s_{ij} = 2k^2,$$

$$(\sigma_1 - \sigma_2)^2 + (\sigma_2 - \sigma_3)^2 + (\sigma_3 - \sigma_1)^2 = 6k^2,$$

$$(\sigma_{11} - \sigma_{22})^2 + (\sigma_{22} - \sigma_{33})^2 + (\sigma_{33} - \sigma_{11})^2 + 6\left(\sigma_{12}^2 + \sigma_{23}^2 + \sigma_{31}^2\right) = 6k^2,$$

where k is the yield stress in simple shear. The reader should show that $Y = \sqrt{3}k$, where Y is the yield stress in simple tension.
The Tresca yield criterion states that yielding of a metal occurs when the maximum shearing stress reaches a critical value, that is, when

$$(\sigma_{\max} - \sigma_{\min}) = 2k,$$

where $\sigma_{\max}$ and $\sigma_{\min}$ are the algebraically greatest and least principal stresses.

3.7 Nominal Stress Tensor

In order to introduce the concept of the nominal stress tensor we first consider simple tension of a straight bar of uniform cross section. The Cauchy or true tensile stress is equal to the tensile force divided by the current cross-sectional area of the bar. However, when the stress is taken as the force divided by the cross section of the bar in its undeformed natural reference configuration, it is said to be the nominal stress. This can be generalized and made more precise by introducing the nominal stress tensor $\boldsymbol{\Sigma}$ with components Σ_{Ki}, where $\Sigma_{Ki}dA$ is the ith component of force currently acting on the plane element, which, in the undeformed natural reference configuration, is normal to the Kth axis and is of infinitesimal area dA. The vector form of the element dA is $d\mathbf{A} = \boldsymbol{N}dA$, where $\boldsymbol{N} = N_K\boldsymbol{E}_k$ is the unit normal to dA and the $\boldsymbol{E}_K$ are the unit base vectors. Unit normal $\boldsymbol{N}$ and area dA in the reference configuration become unit normal $\boldsymbol{n}$ and area da in the current configuration. The force acting on the element dA is denoted by $\boldsymbol{T}(\boldsymbol{N})dA$, where $\boldsymbol{T}(\boldsymbol{N})$ is the nominal stress vector acting on dA and is given by

$$T_i = \Sigma_{Ki}N_K, \; \boldsymbol{T} = \boldsymbol{\Sigma}^{\boldsymbol{T}}\boldsymbol{N}. \tag{3.30}$$

Equation (3.30) may be obtained in a similar manner to equation (3.10) with the tetrahedron in Figure 3.3 taken in the reference configuration. This proof is left as an exercise for the reader. It may be deduced from equation (3.30) that $\boldsymbol{\Sigma}$ is related to both the reference configuration and the spatial configuration and it is called a second-order two-point tensor. The transpose of $\boldsymbol{\Sigma}$ is known as the first Piola-Kirchhoff stress tensor.

The relation between the Cauchy stress and the nominal stress $\boldsymbol{\Sigma}$ is obtained as follows. It follows from equation (3.10) and

$$t_i da = T_i dA, \quad \boldsymbol{t} da = \boldsymbol{T} dA$$

that

$$\sigma_{ij} n_j da = \Sigma_{Ki} N_K dA,$$

and substituting the result (exercise 2, chapter 2)

$$n_i da = \frac{\rho_0}{\rho} \frac{\partial X_K}{\partial x_i} N_K dA,$$

where ρ_0 and ρ are the densities in the reference and spatial configurations, respectively, gives

$$\Sigma_{Ki} = \frac{\rho_0}{\rho} \frac{\partial X_K}{\partial x_j} \sigma_{ji}, \quad \boldsymbol{\Sigma} = \frac{\rho_0}{\rho} \boldsymbol{F}^{-1} \boldsymbol{\sigma}. \tag{3.31}$$

The nominal stress tensor may be expressed in terms of its Cartesian components as

$$\boldsymbol{\Sigma} = \Sigma_{Ki} \boldsymbol{E}_K \otimes \mathbf{e_i}.$$

The form of equation (3.3) in terms of integrals over the reference volume V and surface ∂V is

$$\begin{aligned} \int_{\partial V} T_i dA + \int_{V} \rho_0 f_{0i} dV &= \int_{V} \rho_0 \frac{\partial v_{0i}(X_K, t)}{\partial t} dV, \\ \int_{\partial V} \boldsymbol{T} dA + \int_{V} \rho_0 \boldsymbol{f_0} dV &= \int_{V} \rho_0 \frac{\partial \boldsymbol{v}_o(\boldsymbol{X}, t)}{\partial t} dV, \end{aligned} \tag{3.32}$$

where $\boldsymbol{T} = \boldsymbol{\Sigma}^T \boldsymbol{N}$, $\boldsymbol{f}_0(\boldsymbol{X}, t) = \boldsymbol{f}(\boldsymbol{x}(\boldsymbol{X}, t), t)$ and $\boldsymbol{v}_o = \boldsymbol{v}(\boldsymbol{x}(\boldsymbol{X}, t), t)$. Substituting equation (3.30) in equation (3.32), applying the divergence theorem, and noting that equation (3.32) applies to any part of the body gives the equation of motion:

$$\begin{aligned} \frac{\partial \Sigma_{Ki}}{\partial X_K}(\boldsymbol{X},t) + \rho_0(\boldsymbol{X}) f_{0i}(\boldsymbol{X},t) &= \rho_0 \frac{\partial v_{0i}(\boldsymbol{X},t)}{\partial t}, \\ \mathrm{Div}\boldsymbol{\Sigma}^T(\boldsymbol{X},t) + \rho_0(\boldsymbol{X}) f_0(\boldsymbol{X},t) &= \rho_0(\boldsymbol{X}) \frac{\partial \boldsymbol{v}_0(\boldsymbol{X},t)}{\partial t}, \end{aligned} \tag{3.33}$$

where the notation Div is used to denote the divergence operation based on the operator,

$$\nabla_X \equiv \frac{\partial(\ )}{\partial X_K}\boldsymbol{E_K},$$

but each term of equation (3.33) is a vector referred to the spatial configuration.

3.8 Second Piola-Kirchhoff Stress Tensor

The second Piola-Kirchhoff stress tensor,

$$\mathbf{S} = \mathrm{S}_{KL}\boldsymbol{E}_K \otimes \boldsymbol{E}_L,$$

is a symmetric second-order tensor, based on the reference configuration. It is related to the Cauchy and nominal stress tensors as follows:

$$\mathrm{S}_{KL} = J\frac{\partial X_K}{\partial x_i}\frac{\partial X_L}{\partial x_j}\sigma_{ij}, \quad \mathbf{S} = J\boldsymbol{F}^{-1}\boldsymbol{\sigma}\boldsymbol{F}^{-1^T} \tag{3.34}$$

$$\mathrm{S}_{KL} = \frac{\partial X_K}{\partial x_i}\Sigma_{Li}, \quad \mathbf{S} = \boldsymbol{F}^{-1}\boldsymbol{\Sigma}^T = \boldsymbol{\Sigma}\boldsymbol{F}^{-1^T}. \tag{3.35}$$

The tensor $\boldsymbol{S}$ unlike the tensor $\boldsymbol{\Sigma}$ is symmetric, and this may be advantageous in some applications, but it cannot be directly related to the stress vector acting on a plane element. It is a particular case of the class of Lagrangian tensors.

3.9 Other Stress Tensors

The following stress tensor is obtained from the nominal stress tensor as follows:

$$\hat{\mathbf{S}} = \Sigma R = \Sigma_{Ki} R_{iL} \boldsymbol{E}_K \otimes \boldsymbol{E}_L, \tag{3.36}$$

and its symmetric part

$$\mathbf{S}^B = \frac{1}{2}\left(\Sigma \boldsymbol{R} + \boldsymbol{R}^T \Sigma^T\right), \quad S^B_{KL} = \frac{1}{2}(\Sigma_{Ki} R_{iL} + R_{iK} \Sigma_{Li}) \tag{3.37}$$

is known as the Biot stress tensor. Equation (3.36) gives rise to the polar decomposition $\Sigma = \hat{\boldsymbol{S}} \boldsymbol{R}^T$, which is analogous to the polar decomposition of the deformation gradient tensor $\boldsymbol{F}$. However, $\hat{\mathbf{S}}$ is not unique since the deformation due to a nominal stress Σ may not be unique [3]. Using equation (3.35) and the polar decomposition $\boldsymbol{F} = \boldsymbol{R}\boldsymbol{U}$, equation (3.37) can be put in the form

$$\boldsymbol{S}^B = \frac{1}{2}(\boldsymbol{S}\boldsymbol{U} + \boldsymbol{U}\boldsymbol{S}), \quad S^B_{KL} = \frac{1}{2}(S_{KL} U_{LJ} + U_{KL} S_{LJ}). \tag{3.38}$$

The Biot stress tensor is useful when certain problems of isotropic elasticity are considered. For isotropic elasticity the tensors $\boldsymbol{S}$ and $\boldsymbol{U}$ are coaxial so that the product is commutative. It then follows that

$$\boldsymbol{S}^B = \boldsymbol{U}\boldsymbol{S}, \tag{3.39}$$

which indicates that $\boldsymbol{S}^B$ is coaxial with $\boldsymbol{U}$ and $\boldsymbol{S}$.

The relations $\Sigma = \boldsymbol{S}\boldsymbol{F}^T$ and $\Sigma^{\mathrm{T}} = \boldsymbol{F}\boldsymbol{S}$, obtained from equation (3.35), and $\boldsymbol{F}^T \boldsymbol{F} = \boldsymbol{U}^2$ give $\Sigma\Sigma^T = \boldsymbol{S}\boldsymbol{U}^2\boldsymbol{S}$. It follows from this result, equation (3.39), and the symmetry of $\mathbf{S}$ and $\mathbf{U}$ that

$$\boldsymbol{S}^B = \sqrt{\Sigma\Sigma^T}. \tag{3.40}$$

It should be noted that equations (3.39) and (3.40) are valid only for isotropic elastic media. It may be shown that $\mathbf{\Sigma\Sigma}^T$ is positive definite or positive semidefinite if $\mathbf{\Sigma}$ is nonsingular or singular, respectively. If $\mathbf{\Sigma\Sigma}^T$ is positive definite its principal values $z^{(1)}, z^{(2)}$, and $z^{(3)}$ are positive. If $\mathbf{\Sigma\Sigma}^T$ is positive semidefinite, its principal values are non-negative. The principal values of $\boldsymbol{S}^B = \sqrt{\mathbf{\Sigma\Sigma}^T}$ are the positive or negative square roots of $z^{(1)}$, $z^{(2)}$, and $z^{(3)}$. It is evident that a given $\mathbf{\Sigma}$, does not, in general, determine a unique $\boldsymbol{S}^B$, and this is discussed in detail by Ogden [3].

It is easily shown that $\det[\mathbf{\Sigma}] = \det\left[\mathbf{S}^B\right]$ and that $\Sigma_{Ki}\delta_{iL} = S_{KL}$ if $\boldsymbol{R} = \boldsymbol{I}$.

EXERCISES

3.1. Given the state of true stress with the following components

$$\begin{bmatrix} 3 & 1 & 3 \\ 1 & -2 & -2 \\ 3 & -2 & 5 \end{bmatrix}$$

Find

(a) the principal Cauchy stresses,

(b) the octahedral shearing stress,

(c) the direction cosines of the normals to the octahedral planes,

(d) the normal stress, shearing stress, and direction of the shearing stress acting on the plane with unit normal

$$\mathbf{n} = \frac{1}{2}\mathbf{e}_1 + \frac{1}{2}\mathbf{e}_2 + \frac{1}{\sqrt{2}}\mathbf{e}_3 .$$

3.2. Show that if the Cauchy stress tensor is singular, that is, $\det[\sigma_{ij}] = 0$, one of the principal stresses is zero.

3.3. Prove that if the Cauchy stress tensor $\boldsymbol{\sigma}$ and left Cauchy-Green strain tensor $\boldsymbol{B}$ are coaxial, the second Piola-Kirchhoff stress tensor $\boldsymbol{S}$ and the right Cauchy-Green strain tensor $\boldsymbol{C}$ are also coaxial.

3.4. Consider the following state of simple shear

$$x_1 = X_1 + KX_2, \quad x_2 = X_2, \quad x_3 = X_3.$$

The components of the Cauchy stress tensor required to produce this deformation are

$$\begin{bmatrix} -\mathrm{p} & \alpha & 0 \\ \alpha & -\mathrm{p}_1 & 0 \\ 0 & 0 & \mathrm{p}_2 \end{bmatrix}.$$

Find the matrix of components of the first and second Piola-Kirchhoff stress tensors.

3.5. Prove that if a body of volume V and surface A is in equilibrium under surface traction $\boldsymbol{t}(\boldsymbol{x}), \boldsymbol{x} \in A$ where $\boldsymbol{x}$ is the position vector of a material point with respect to a fixed origin the mean hydrostatic part of the stress tensor, σ_m, is given by,

$$\sigma_m = \frac{1}{3V} \int_{\partial V} \boldsymbol{t} \cdot \boldsymbol{X} \mathrm{da}.$$

3.6. Let $\boldsymbol{m}$ and $\boldsymbol{n}$ be unit vectors normal to surface elements $\Delta \mathrm{a}_m$ and $\Delta \mathrm{a}_n$ at a point, and let $\boldsymbol{t}(\boldsymbol{m})$ and $\boldsymbol{t}(\boldsymbol{n})$ be the stress vectors acting on these plane elements. Show that

$$\boldsymbol{n} \cdot t(\boldsymbol{m}) = \boldsymbol{m} \cdot \boldsymbol{t}(\boldsymbol{n}).$$

3.7. Consider a state of plane stress with $\sigma_{33} = \sigma_{32} = \sigma_{31} = 0$. Let

$$\mathrm{N}_{(1)}, \mathrm{N}_{(2)}, \text{ and } \mathrm{N}_{(3)}$$

be the normal stress on plane elements that are normal to the x_3 plane and form angles of 120° with each other. Find the principal stress and the matrix of components of the stress tensor if the normal stress $\mathrm{N}_{(1)}$ is in the x_1 direction and the direction of $\mathrm{N}_{(2)}$ is in the second quadrant of the $0x_1x_2$ plane.

3.8. If rectangular Cartesian axes $O\hat{x}_i$, $i \in \{1, 2, 3\}$ are obtained from Ox_i by rotating about Ox_3 through angle α, using the right-hand rule, show that

$$\hat{\sigma}_{11} + \hat{\sigma}_{22} = \sigma_{11} + \sigma_{22},$$

$$\hat{\sigma}_{11} - \hat{\sigma}_{22} + 2i\hat{\sigma}_{12} = \mathrm{e}^{-2i\alpha}(\sigma_{11} - \sigma_{22} + 2i\sigma_{12}),$$

$$\hat{\sigma}_{13} + i\hat{\sigma}_{23} = \mathrm{e}^{-i\alpha}(\sigma_{13} + i\sigma_{23}),$$

where $\hat{\sigma}_{ij}$ and σ_{ij} are the components of the Cauchy stress with respect to $O\hat{x}_i$ and Ox_i, respectively.

3.9. A body of current volume V is in equilibrium under the action of surface traction $\boldsymbol{\sigma}\mathbf{n}$ and body force $\mathbf{f}$. Show that the mean stress $\bar{\boldsymbol{\sigma}} = \frac{1}{\mathrm{V}} \int_{\mathrm{V}} \boldsymbol{\sigma} d\mathrm{V}$ is given by

$$\bar{\boldsymbol{\sigma}} = \frac{1}{\mathrm{V}} \left\{ \int_{\partial \mathrm{v}} \left(\boldsymbol{\sigma}^{\mathrm{T}} \boldsymbol{n} \otimes \boldsymbol{x} \right) \mathrm{da} + \int_{\mathrm{v}} \boldsymbol{f} \otimes \boldsymbol{x} \mathrm{dv} \right\}.$$

3.10. Prove that

$$\boldsymbol{S}\dot{\boldsymbol{C}} = 2\boldsymbol{\Sigma}\dot{\boldsymbol{F}}.$$

REFERENCES

1. Pearson, C.E. (1959). Theoretical Elasticity. Harvard University Press.
2. Hill, R. (1950). The Mathematical Theory of Plasticity. Oxford University Press.
3. Ogden, R.W. (1984). Non-Linear Elastic Deformations. Ellis Harwood.

4 Work, Energy, and Entropy Considerations

This chapter is concerned with some aspects of work, energy, and contiuum thermodynamics. The treatment is general and reference is not made to any particular material. Notation for deformation and stress variables is the same as in chapters 2 and 3, where the lower- and uppercase suffixes refer to the spatial and reference configurations, respectively. Most of the equations in this chapter are given in both suffix and symbolic notation.

4.1 Stress Power

Stress power is the rate of work done by the stresses and is a purely mechanical quantity, that is, obtained from the rate of work of the body forces and surface tractions acting on a material body and does not involve any other forms of energy transfer such as heating. It is a specific quantity per unit volume or per unit mass and can be expressed in terms of either the Cauchy, nominal, or second Piola-Kirchhoff stress and the respective conjugate rate variable.

The rate of work of the body forces and surface tractions acting on a material body of volume v and surface a in the spatial configuration is given by

$$\dot{W} = \int_{\partial \mathrm{v}} t_i v_i d\mathrm{a} + \int_{\mathrm{v}} \rho f_i v_i d\mathrm{v}, \qquad (4.1)$$
$$\dot{W} = \int_{\partial \mathrm{v}} \boldsymbol{t} \cdot \boldsymbol{v} d\mathrm{a} + \int_{\mathrm{v}} \rho \boldsymbol{f} \cdot \boldsymbol{v} d\mathrm{v},$$

where ρ is the density in the spatial configuration and the vectors $\boldsymbol{t}$, $\boldsymbol{v}$, and $\boldsymbol{f}$ are the surface traction, velocity, and body force per unit mass, respectively, and are functions of $\boldsymbol{x}$ and t. Substituting equation (3.13),

$$t_i = \sigma_{ij} n_j, \quad \boldsymbol{t} = \boldsymbol{\sigma n}, \tag{4.2}$$

where $\boldsymbol{\sigma}$ is the Cauchy stress tensor and $\boldsymbol{n}$ is the outward unit normal to the surface, in equation (4.1) and applying the divergence theorem gives

$$\begin{aligned} \dot{W} &= \int_{\mathrm{V}} \left\{ v_i \left[\frac{\partial \sigma_{ij}}{\partial x_j} + \rho f_i \right] + \sigma_{ij} L_{ij} \right\} d\mathrm{v} \\ \dot{W} &= \int_{\mathrm{V}} \{ \boldsymbol{v} \cdot [\mathrm{div}\boldsymbol{\sigma} + \rho \boldsymbol{f}] + \mathrm{tr}(\boldsymbol{\sigma L}) \} d\mathrm{v} \end{aligned} \tag{4.3}$$

where $L_{ij} = \partial v_i / \partial x_j$ are the components of the velocity gradient tensor $\nabla \otimes \boldsymbol{v}$. With the use of the equation of motion

$$\begin{aligned} \frac{\partial \sigma_{ij}}{\partial x_j} + \rho f_i &= \rho \frac{dv_i}{dt}, \\ \mathrm{div}\boldsymbol{\sigma} + \rho \boldsymbol{f} &= \rho \frac{d\boldsymbol{v}}{dt}, \end{aligned} \tag{4.4}$$

equation (4.3) becomes

$$\begin{aligned} \dot{W} &= \frac{d}{dt} \int_{\mathrm{V}} \rho \frac{v_i v_i}{2} d\mathrm{v} + \int_{\mathrm{V}} \sigma_{ij} L_{ij} d\mathrm{v}, \\ \dot{W} &= \frac{d}{dt} \int_{\mathrm{V}} \rho \frac{\boldsymbol{v} \cdot \boldsymbol{v}}{2} d\mathrm{v} + \int_{\mathrm{V}} \mathrm{tr}(\boldsymbol{\sigma L}) d\mathrm{v}. \end{aligned} \tag{4.5}$$

The first term on the right-hand side of equation (4.5) is the rate of change of the kinetic energy of the body, and the second term is integral over the volume of the specific stress power per unit volume of the spatial configuration of the body. It follows that the specific stress power per unit mass is

$$P = \frac{1}{\rho} \sigma_{ij} L_{ij} = \frac{1}{\rho} \mathrm{tr}(\boldsymbol{\sigma L}). \tag{4.6}$$

Decomposition of the velocity gradient tensor into symmetric and antisymmetric parts is given by

$$L_{ij} = d_{ij} + \omega_{ij}, \quad \boldsymbol{L} = \boldsymbol{D} + \boldsymbol{W}, \tag{4.7}$$

where $\boldsymbol{D} = (\boldsymbol{L} + \boldsymbol{L}^T)/2$ and $\boldsymbol{W} = (\boldsymbol{L} - \boldsymbol{L}^T)/2$ are the rate of deformation and spin tensors, respectively. Substitution of equation (4.7) in equation (4.6) gives

$$P = \frac{1}{\rho}\sigma_{ij}D_{ij} = \frac{1}{\rho}\mathrm{tr}(\boldsymbol{\sigma D}), \tag{4.8}$$

since $\omega_{ij}\sigma_{ij} = \mathrm{tr}(\boldsymbol{W\sigma}) = 0$.

The specific stress power may also be expressed in terms of the first and second Piola-Kirchhoff stresses and the respective conjugate rate variables. In this treatise, the nominal stress tensor $\boldsymbol{\Sigma}$, which is the transpose of the first Piola-Kirchhoff stress, is used. Component Σ_{Kj}, of $\boldsymbol{\Sigma}$, as noted in chapter 3, is the jth component of the stress vector acting on the plane element, referred to the reference configuration and whose normal is in the Kth direction. In order to express P in terms of the nominal stress tensor, $\boldsymbol{\Sigma}$, equation (4.1) is replaced by

$$\begin{aligned} \dot{W} &= \int_{\partial V} T_i v_{0i} dA + \int_{V} \rho_0 f_{0i} v_{0i} dV, \\ \dot{W} &= \int_{\partial V} \boldsymbol{T} \cdot \boldsymbol{v_0} dA + \int_{V} \rho_0 \boldsymbol{f_0} \cdot \boldsymbol{v_0} dV, \end{aligned} \tag{4.9}$$

where

$$(\rho_0, v_0, \boldsymbol{f}_0) = (\rho, \boldsymbol{v}, \boldsymbol{f})(\boldsymbol{x}(\boldsymbol{X}, t), t)$$

and

$$T_i = \Sigma_{Ki} N_K, \; \boldsymbol{T} = \boldsymbol{\Sigma}^T \boldsymbol{N} \tag{4.10}$$

is the nominal surface traction and ρ_0, V, A, and $\boldsymbol{N}$ denote the density, volume, surface, and outward unit normal to the surface, respectively, of a material body in the reference configuration. In equation (4.9) the integrands $\boldsymbol{T} \cdot \boldsymbol{v}_0$ and $\rho_0 \boldsymbol{f}_0 \cdot \boldsymbol{v}_0$ are the rate of work of the surface tractions per unit area of the surface of the reference configuration and the rate of work of the body force per unit volume of the reference configuration, respectively. Substituting equation (4.10) in equation (4.9) and applying the divergence theorem gives

$$\begin{aligned} \dot{W} &= \int_{\mathrm{V}} \left\{ v_i \left\{ \frac{\partial \Sigma_{Ki}}{\partial X_K} + \rho_0 f_{0i} \right\} + \Sigma_{Ki} \dot{F}_{iK} \right\} d\mathrm{V}, \\ \dot{W} &= \int_{\mathrm{V}} \left\{ \boldsymbol{v}. \left\{ \mathrm{Div} \boldsymbol{\Sigma}^{\mathrm{T}} + \rho_0 \boldsymbol{f}_0 \right\} + \mathrm{tr} \left(\Sigma \dot{\boldsymbol{F}} \right) \right\} d\mathrm{V}, \end{aligned} \tag{4.11}$$

where $F_{iK} = \partial x_i / \partial X_K$, are the components of the deformation gradient tensor and $\dot{F}_{iK} = L_{ij} F_{jK}$, $\dot{\boldsymbol{F}} = \boldsymbol{L}\boldsymbol{F}$. With the use of the equations of motion

$$\begin{aligned} \frac{\partial \Sigma_{Ki}}{\partial X_K} + \rho_0 f_{0i} &= \rho_0 \frac{\partial v_{0i}}{\partial t}, \\ \mathrm{Div} \Sigma^T + \rho_0 \boldsymbol{f}_0 &= \rho_0 \frac{\partial \boldsymbol{v}_0}{\partial t}, \end{aligned} \tag{4.12}$$

equation (4.11) becomes

$$\begin{aligned} \dot{W} &= \frac{\partial}{\partial t} \int_{\mathrm{V}} \rho_0 \frac{v_{0i} v_{0i}}{2} d\mathrm{V} + \int_{\mathrm{V}} \Sigma_{Ki} \dot{F}_{iK} d\mathrm{V}, \\ \dot{W} &= \frac{\partial}{\partial t} \int_{\mathrm{V}} \rho_0 \frac{\boldsymbol{v}_0 \cdot \boldsymbol{v}_0}{2} d\mathrm{V} + \int_{\mathrm{V}} \mathrm{tr} \left(\boldsymbol{\Sigma} \dot{\boldsymbol{F}} \right) d\mathrm{V}. \end{aligned} \tag{4.13}$$

It follows from equation (4.13) that the specific stress power (per unit mass) is given by

$$P = \frac{1}{\rho_0} \left(\Sigma_{Ki} \dot{F}_{iK} \right) = \frac{1}{\rho_0} \mathrm{tr} \left(\boldsymbol{\Sigma} \dot{\boldsymbol{F}} \right). \tag{4.14}$$

The stress power in terms of the second Piola-Kirchhoff stress $\boldsymbol{S}$ and $\dot{\boldsymbol{E}}$, the Lagrangian strain rate, cannot be determined directly in the same manner as equations (4.8) and (4.14) but can be determined from equation (4.8) or (4.14). Substitution of

$$\sigma_{ij} = \frac{\rho}{\rho_0} S_{KL} \frac{\partial x_i}{\partial X_K} \frac{\partial x_j}{\partial X_L}, \qquad \boldsymbol{\sigma} = \frac{\rho}{\rho_0} \boldsymbol{F S F}^T,$$

obtained from equation (3.34), and

$$d_{ij} = \dot{E}_{MN} \frac{\partial X_M}{\partial x_i} \frac{\partial X_N}{\partial x_j}, \qquad \boldsymbol{D} = \boldsymbol{F}^{-T} \dot{\boldsymbol{E}} \boldsymbol{F}^{-1},$$

obtaind from equation (2.29), in equation (4.8) gives

$$P = \frac{1}{\rho_0} S_{KL} \dot{E}_{KL} = \frac{1}{\rho_0} \mathrm{tr}(\boldsymbol{S}\dot{\boldsymbol{E}}) = \frac{1}{2\rho_0} \mathrm{tr}(\boldsymbol{S}\dot{\boldsymbol{C}}). \tag{4.15}$$

An alternative determination of equation (4.15) is as follows. Substitution of

$$\Sigma_{Ki} = S_{KL} F_{iL}, \quad \boldsymbol{\Sigma} = \boldsymbol{S F}^{\boldsymbol{T}}$$

in equation (4.14) gives

$$P = \frac{1}{\rho_0} \mathrm{tr}\left(\boldsymbol{S F}^T \dot{\boldsymbol{F}}\right). \tag{4.16}$$

It follows from equation (2.60) that

$$\dot{E}_{KL} = (\dot{F}_{iK} F_{iL} + F_{iK} \dot{F}_{iL})/2, \quad \dot{\mathrm{E}} = (\dot{\boldsymbol{F}}^{\boldsymbol{T}} \boldsymbol{F} + \boldsymbol{F}^{\boldsymbol{T}} \dot{\boldsymbol{F}})\big/2. \tag{4.17}$$

Substitution of equation (4.17) in equation (4.16) gives equation (4.15) since

$$\mathrm{tr}\left(\boldsymbol{S} \dot{\boldsymbol{F}}^T \boldsymbol{F}\right) = \frac{1}{2} \mathrm{tr}\left(\boldsymbol{S}\left(\dot{\boldsymbol{F}}^T \boldsymbol{F} + \boldsymbol{F}^T \dot{\boldsymbol{F}}\right)\right) = \mathrm{tr}(\boldsymbol{S}\dot{\boldsymbol{E}}).$$

4.2 Principle of Virtual Work

The principle of virtual work applied to static problems involves the spatial form of the equations of equilibrium,

$$\frac{\partial \sigma_{ij}}{dx_j} + \rho f_i = 0, \qquad \text{div}\boldsymbol{\sigma} + \rho \boldsymbol{f} = 0, \tag{4.18}$$

or the referential form,

$$\frac{\partial \Sigma_{Ki}}{\partial X_K} + \rho_0 f_{0i} = 0, \qquad \text{Div}\boldsymbol{\Sigma}^T + \rho_0 \boldsymbol{f_0} = 0, \tag{4.19}$$

rather than the equations of motion.

In order to introduce the principle of virtual work, a boundary value problem is posed and a statically admissible stress field and a kinematically admissible virtual displacement field are defined. A statically admissible stress field is a stress field that satisfies the equilibrium equations and the stress boundary conditions. Stress discontinuities occur in certain static quasi-static problems, for example, at the contact surface of two component cylinders of a compound cylinder with shrink fit. Equilibrium requires that the normal component, $t_i n_i = \boldsymbol{t} \cdot \boldsymbol{n}$, of the stress vector acting on the discontinuity surface must be continuous, where $\boldsymbol{n}$ is the unit normal on one side of the discontinuity surface.

A kinematically admissible virtual incremental displacement field, $\delta \boldsymbol{u}(\boldsymbol{x})$, is an imagined displacement that is assumed to be infinitessimal, continuous, and vanishes on the part of the boundary $\partial \mathrm{v}_u$ upon which the displacements are prescribed. It should be emphasized that the stress field remains unchanged during the application of $\delta \boldsymbol{u}(\boldsymbol{x})$. Since it is assumed that $\delta \boldsymbol{u}(\boldsymbol{x})$ is infinitesimal, it is consistent with the concept of virtual displacements in classical mechanics. The distinction between $\delta \boldsymbol{u}(\boldsymbol{x})$ and $d\boldsymbol{u}(\boldsymbol{x})$, that is, an actual increment of displacement, is worth noting. An actual infinitesimal incremental displacement field could be regarded as a virtual incremental displacement field, but the converse is not necessarily true.

First we consider a boundary value problem referred to the spatial configuration. Then, the body force field acting on a body could be prescribed along with the surface tractions on part $\partial \mathrm{v}_t$ of the surface and the displacements on the remainder $\partial \mathrm{v}_u$ of the surface. It is also possible that the normal component of stress and the tangential component of displacement or vice versa could be prescribed on a part of the surface. Malvern [1] has considered further combinations of traction and displacement components prescribed on part of the surface. It is easy to modify the theory to consider these cases; however, we will assume the tractions and displacements are prescribed on disjoint parts of the surface so that $\partial \mathrm{v}_t \cup \partial \mathrm{v}_u = \partial \mathrm{v}$ and $\partial \mathrm{v}_t \cap \partial \mathrm{v}_u = 0$ where the symbols $\cup$ and $\cap$ denote the union of the two surfaces and the intersection of the two surfaces, respectively.

The virtual work of the surface tractions and body force, referred to the spatial configuration for static problems, is

$$\delta W = \int_{\partial \mathrm{v}_\tau} t_i \delta u_i d\mathrm{a} + \int_{\mathrm{v}} \rho f_i \delta u_i d\mathrm{v}, \tag{4.20}$$

where $\delta \boldsymbol{u} = \boldsymbol{0}$ on the part $\partial \mathrm{v}_u$ of the surface. With the use of equation (3.10) and the divergence theorem (4.20) becomes

$$\begin{aligned} \delta W &= \int_{\mathrm{v}} \left\{ \frac{\partial (\sigma_{ij} \delta u_j)}{\partial x_i} + \rho f_i \delta u_i \right\} d\mathrm{v} \\ &= \int_{\mathrm{v}} \left\{ \frac{\partial \sigma_{ij}}{\partial x_i} \delta u_j + \sigma_{ij} \frac{\partial \delta u_j}{\partial x_i} + \rho f_i \delta u_i \right\} d\mathrm{v}. \end{aligned} \tag{4.21}$$

Then, with the use of the equation of equilibrium (4.18), (4.20), (4.21), and $\sigma_{ij} \partial \delta u_i / \partial x_j = \sigma_{ij} \delta e_{ij}$,

$$\begin{aligned} &\int_{\partial \mathrm{v}_\tau} t_i \delta u_i d\mathrm{a} + \int_{\mathrm{v}} \rho f_i \delta u_i d\mathrm{v} = \int_{\mathrm{v}} \sigma_{ij} \delta e_{ii} d\mathrm{v}, \\ &\int_{\partial \mathrm{v}_\tau} \boldsymbol{t} \cdot \delta \boldsymbol{u}_i d\mathrm{a} + \int_{\mathrm{v}} \rho \boldsymbol{f} \cdot \delta \boldsymbol{u} d\mathrm{v} = \int_{\mathrm{v}} \mathrm{tr}(\boldsymbol{\sigma} \delta \boldsymbol{e}) d\mathrm{v}, \end{aligned} \tag{4.22}$$

where

$$\delta e_{ij} = \frac{1}{2}\left(\frac{\partial \delta u_i}{\partial x_j} + \frac{\partial \delta u_j}{\partial x_i}\right), \quad \delta \boldsymbol{e} = \frac{1}{2}\left(\nabla \otimes \delta \boldsymbol{u} + (\nabla \otimes \delta \boldsymbol{u})^T\right), \tag{4.23}$$

which is a virtual infinitesimal strain increment, based on the current configuration. Equation (4.22) is a form of the principle of virtual work and indicates that, for any admissible virtual displacement field, the virtual work (4.20) of the external forces is equal to the virtual work of the internal forces. It is usually assumed that the δ variation and the gradient operator ∇ commute, and with this assumption equation (4.23) is equivalent to

$$\delta \boldsymbol{e} = \delta \frac{1}{2}\left((\nabla \otimes \boldsymbol{u}) + (\nabla \otimes \boldsymbol{u})^T\right), \tag{4.24}$$

which is a virtual increment of the infinitessimal strain tensor. A difficulty arises when a problem involving finite deformation is considered, then equation (4.23) is appropriate but not equation (4.24).

The form of the principle of virtual work, for a statically admissible stress field in terms of the nominal stress $\boldsymbol{\Sigma}$ and a virtual displacement field, $\delta \widehat{\boldsymbol{u}}(\boldsymbol{X}) = \delta \boldsymbol{u}(\boldsymbol{x}(\boldsymbol{X}))$ is

$$\begin{gathered} \int\limits_{\partial \mathrm{V}_T} T_i \delta \widehat{u}_i dA + \int\limits_{\mathrm{V}} \rho_0 f_{0i} \delta \widehat{u}_i d\mathrm{V} = \int\limits_{\mathrm{V}} \Sigma_{Ki} \delta F_{iK} d\mathrm{V}, \\ \int\limits_{\partial \mathrm{V}_T} \boldsymbol{T} \cdot \delta \widehat{\boldsymbol{u}} dA + \int\limits_{\mathrm{V}} \rho_0 \boldsymbol{f}_0 \cdot \delta \widehat{\boldsymbol{u}} d\mathrm{V} = \int\limits_{\mathrm{V}} tr(\Sigma \delta \boldsymbol{F}) d\mathrm{V}, \end{gathered} \tag{4.25}$$

referred to the reference configuration, where V is the volume in the reference configuration and $\partial \mathrm{V}_T$ is the part of the reference surface $\partial \mathrm{V}$ upon which the surface traction given by equation (4.10) is prescribed. The proof of equation (4.25) is similar to that for equation (4.22) with T_i eliminated using equation (4.10) followed by use of the divergence theorem and equation (4.19).

It is evident from equations (4.22) and (4.25) that, for a static problem, the virtual work of the applied forces, that is, the surface tractions and body forces, is equal to virtual work of the internal forces, that is, the virtual work of the internal stresses.

Another form of the principle of virtual work for static or quasi-static processes arises when we consider a statically admissible stress field $\boldsymbol{\sigma}(\boldsymbol{x})$ along with a kinematically admissible velocity field $v(\boldsymbol{x})$ in a region v with surface ∂v of a material body. A kinematically admissible velocity field satisfies the velocity boundary conditions on the part of the boundary upon which the velocity is prescribed and any kinematical constraints such as incompressibility. We then obtain the result for quasi-static processes

$$\begin{aligned}
\int_{\delta v_t} t_i v_i da + \int_{v} \rho v_i f_i dv &= \int_{v} \sigma_{ij} d_{ij} dv, \\
\int_{\delta v_t} \boldsymbol{t}.\boldsymbol{v} da + \int_{v} \rho \boldsymbol{v} \cdot \boldsymbol{f} dv &= \int_{v} \mathrm{tr}(\boldsymbol{\sigma D}) dv,
\end{aligned} \tag{4.26}$$

where $\boldsymbol{D}$ is the rate of deformation tensor obtained from the kinematically admissible velocity field $\boldsymbol{v}(\boldsymbol{x})$. It is important to note that there need be no cause and effect relationship between the statically admissible stress field and the kinematically admissible velocity field in equation (4.26). In some applications it is convenient to consider velocities and rates of deformation with respect to a monotonically varying parameter of the problem rather than time. This is often the case for quasi-static problems in the theory of plasticity. In the theory of rigid perfectly plastic solids volume preserving discontinuities may occur in kinematically admissible velocity fields for certain problems.

The principle of virtual work may also be applied to dynamic problems and, referred to the spatial configuration, is then given by

$$\begin{aligned}
\int_{\partial v_t} t_i \delta u_i da + \int_{v} \rho f_i \delta u_i dv - \int_{v} \rho \dot{v}_i \delta u_i dv &= \int_{v} \sigma_{ij} \delta e_{ij} dv, \\
\int_{\partial v_t} \boldsymbol{t} \cdot \delta \boldsymbol{u} da + \int_{v} \rho \boldsymbol{f} \cdot \delta \boldsymbol{u} dv - \int_{v} \rho \dot{\boldsymbol{v}} \cdot \delta \boldsymbol{u} dv &= \int_{v} tr(\boldsymbol{\sigma} \delta e) dv,
\end{aligned} \tag{4.27}$$

where the virtual displacement field $\delta \boldsymbol{u}(\boldsymbol{x})$ is continuous and vanishes on the part of the boundary upon which the velocity is prescribed and the deformation of the material body is imagined to be frozen when the virtual displacement field is applied. Equation (4.27) shows that the virtual work of the applied forces and inertia forces is equal to the virtual work of the internal stresses. The derivation of equation (4.27), which is the continuum form of d'Alembert's Principle of classical mechanics, is similar to the derivation of equation (4.20) with the equilibrium equation (4.18) replaced by the equation of motion (4.4). As an exercise show that the reference configuration form of equation (4.27) is

$$\int_{\partial V} T_i \delta\hat{u}_i dA + \int_V \rho_0 f_{0i} \delta\hat{u}_i dV - \int_V \rho_0 \dot{v}_{0i} \delta\hat{u}_i dV = \int_V \Sigma_{Ki} \delta F_{iK} dV,$$

$$\int_{\partial V} \boldsymbol{T} \cdot \delta\hat{\boldsymbol{u}} dA + \int_V \rho_0 \boldsymbol{f}_0 \cdot \delta\hat{\boldsymbol{u}} dV - \int_V \rho_0 \dot{\boldsymbol{v}}_0 \cdot \delta\hat{\boldsymbol{u}} dV = \int_V tr(\boldsymbol{\Sigma}\delta\boldsymbol{F}) dV.$$

A detailed discussion of the continuum form of d'Alembert's Principle of classical mechanics is given by Haupt [2].

4.3 Energy Equation and Entropy Inequality

In this section we consider some aspects of the thermodynamics of deformation. Continuum mechanics involves the study of nonequilibrium states of continuous media as well as equilibrium states; consequently, in order to introduce thermodynamical concepts, an extension of classical equilibrium thermodynamics, more accurately described as thermostatics, is required. A simple extension of thermostatics to cover irreversible processes is based on the Principle of Local State [3], which is a plausible hypothesis. This principle is as follows:

> *The local and instantaneous relations between the thermodynamic properties of a continuous system are the same as for a uniform system in equilibrium.*

A treatment of thermodynamics of deformation based on this principle is referred to as the thermodynamics of irreversible processes [4]. Essentially what is implied is that infinitesimal elements of a system in a non-equilibrium state behave locally as if undergoing reversible processes. This gives satisfactory results for most engineering problems.

The energy equation is a postulate that follows from the first law of thermodynamics (conservation of energy) for a closed system (material volume). The rate of work of the surface tractions and body forces plus the rate of heat flow in to the body is equal to the rate of change of the sum of the kinetic energy and internal energy. Gravitational potential energy is neglected since it is not significant in most contiuum mechanics problems, but it could easily be included in the equations that follow. The spatial form of the energy equation for a material volume, $\mathrm{v}(t)$ is

$$\frac{d}{dt}\left(\int_{\mathrm{v}} \rho\left(\frac{v_i v_i}{2} + u\right)\right) d\mathrm{v} = \int_{\mathrm{v}} \rho(r + f_i v_i) d\mathrm{v} + \int_{\partial \mathrm{v}} (t_i v_i - q_i n_i) da,$$
$$\frac{d}{dt}\left(\int_{\mathrm{v}} \rho\left(\frac{\boldsymbol{v}\cdot\boldsymbol{v}}{2} + u\right)\right) d\mathrm{v} = \int_{\mathrm{v}} \rho(r + \boldsymbol{f}\cdot\boldsymbol{v}) d\mathrm{v} + \int_{\partial \mathrm{v}} (\boldsymbol{t}\cdot\boldsymbol{v} - \boldsymbol{q}\cdot\boldsymbol{n}) da, \tag{4.28}$$

where $\boldsymbol{n}$ is the outward unit normal to the surface, $u = u(\boldsymbol{x}, t)$ is the specific internal energy, r is rate of heat supply per unit mass, and $\boldsymbol{q}$ is the heat flux vector. Fourier's law for the spatial configuration,

$$q_i = -k_{ij}\frac{\partial \Theta}{\partial x_j}, \quad \boldsymbol{q} = -\boldsymbol{k}\nabla\Theta, \tag{4.29}$$

is a semiempirical relation that relates $\boldsymbol{q}$ and the temperature gradient. In equation (4.29) $\boldsymbol{k}$ is the heat conduction tensor that is a positive semidefinite second-order tensor for an anisotropic medium. For an isotropic medium $k_{ij} = \delta_{ij}k$, where k is a non-negative scalar. The thermal conductivity tensor is a function of temperature, and the deformation and may also be a function of position for an inhomogeneous body. For many problems that involve suitably small temperature differences and changes, the temperature dependence may be neglected and $\boldsymbol{k}$ assumed constant. It should be noted that equation (4.29) has a defect, when applied to transient problems, since it predicts an infinite thermal wave speed; however, equation (4.29) is adequate for many engineering applications, and it will be assumed for what follows in this text. Modifications of equation (4.29) in order to remove the defect have been extensively researched and [5] is a useful reference on this topic.

Heat production can arise from various external agencies such as radiation, or from Joule heating in electrical conductors. Potential energy is neglected in equation (4.28). Substituting for t_i, from equation (4.2), in equation (4.28), and using the divergence theorem and the equation of motion (4.4) gives

$$\int_{\mathrm{v}} \left(\rho\dot{u} - \rho r - \sigma_{ij}d_{ij} + \frac{\partial q_i}{\partial x_i}\right) d\mathrm{v} = 0,$$
$$\int_{\mathrm{v}} (\rho\dot{u} - \rho r - \mathrm{tr}(\boldsymbol{\sigma D}) + \mathrm{div}\boldsymbol{q}) d\mathrm{v} = 0. \tag{4.30}$$

Since the integrand is assumed to be continuous and equation (4.30) is valid for volume v or any part of v, it follows that

$$\rho\dot{u} - \rho r - \sigma_{ij}d_{ij} + \frac{\partial q_i}{\partial x_i} = 0,$$
$$\rho\dot{u} - \rho r - \mathrm{tr}(\boldsymbol{\sigma D}) + \mathrm{div}\boldsymbol{q} = 0. \tag{4.31}$$

Equation (4.31) is the local form of the energy equation referred to the spatial configuration. The local form of the energy equation can also be referred to the reference configuration. In order to obtain this form we note that

$$Q_K dA_K = q_i d\mathrm{a}_i, \quad \boldsymbol{Q} \cdot d\boldsymbol{A} = \boldsymbol{q} \cdot d\mathbf{a}, \tag{4.32}$$

where $\boldsymbol{Q}$ is the heat flux vector referred to the reference configuration and $d\boldsymbol{A}$ and $d\boldsymbol{a}$ are the vector areas of a plane element in the reference and spatial configurations, respectively. Using equation (4.32) along with

$$d\mathrm{a}_i = J\frac{\partial X_K}{\partial x_i} dA_K, \quad d\mathbf{a} = J\boldsymbol{F}^{-T} d\boldsymbol{A},$$

gives

$$q_i = J^{-1}F_{iK}Q_K, \quad \boldsymbol{q} = J^{-1}\boldsymbol{FQ}. \tag{4.33}$$

The divergence of equation (4.33) and the result

$$\frac{\partial}{\partial x_i}(J^{-1}F_{iK}) = 0, \quad \mathrm{div}(J^{-1}\boldsymbol{F}) = 0,$$

give

$$\frac{\partial q_i}{\partial x_i} = J^{-1}\frac{\partial Q_K}{\partial X_K}, \quad \mathrm{div}\boldsymbol{q} = J^{-1}\mathrm{Div}\boldsymbol{Q}, \tag{4.34}$$

where, as before,

$$J = \frac{\rho_0}{\rho} = \det[\boldsymbol{F}].$$

It then follows from equations (4.8), (4.14), (4.26), and (4.29) that the local form of the energy equation (4.31), referred to the reference configuration, is

$$\begin{aligned} &\rho_0\dot{u}_0 - \rho_0 r_0 - \Sigma_{Ki}\dot{F}_{iK} + \frac{\partial Q_K}{\partial X_K} = 0, \\ &\rho_0\dot{u}_0 - \rho_0 r_0 - \mathrm{tr}(\Sigma\dot{\boldsymbol{F}}) + \mathrm{Div}\boldsymbol{Q} = 0, \end{aligned} \tag{4.35}$$

where $u_0(\boldsymbol{X}, t) = u(\boldsymbol{x}(\boldsymbol{X}, t), t)$ and $r_0(\boldsymbol{X}, t) = r(\boldsymbol{x}(\boldsymbol{X}, t), t)$. Equation (4.35) can also be obtained from the reference form of equation (4.28),

$$\frac{\partial}{\partial t}\int_{\mathrm{V}} \rho_0(u_0 + \frac{v_{0i}v_{0i}}{2})d\mathrm{V} = \int_{\mathrm{V}} \rho_0(r_0 + f_{oi}v_{0i})d\mathrm{V} + \int_{\partial\mathrm{V}} (T_i v_{0i} - Q_k N_K)dA.$$

In addition to the energy equation we postulate an entropy production relation, that follows from the second law of thermodynamics,

$$\begin{aligned} &\frac{d}{dt}\int_{\mathrm{v}} \rho s d\mathrm{v} - \int_{\mathrm{v}} \frac{\rho r}{\Theta} d\mathrm{v} + \int_{\partial\mathrm{v}} \frac{q_i n_i}{\Theta} d\mathrm{a} \geqslant 0, \\ &\frac{d}{dt}\int_{\mathrm{v}} \rho s d\mathrm{v} - \int_{\mathrm{v}} \frac{\rho r}{\Theta} d\mathrm{v} + \int_{\partial\mathrm{v}} \frac{\boldsymbol{q}\cdot\boldsymbol{n}}{\Theta} d\mathrm{a} \geqslant 0, \end{aligned} \tag{4.36}$$

referred to the spatial configuration, where $s = s(x, t)$ is the specific entropy and Θ is the absolute temperature. Equation (4.36) is known as the Clausius-Duhem inequality. If $\boldsymbol{q} = \boldsymbol{0}$ on ∂v and $r = 0$, the material volume in equation (4.36) is thermally isolated, and

$$\frac{d}{dt}\int_{\mathrm{v}} \rho s d\mathrm{v} = \int_{\mathrm{v}} \rho \dot{s} d\mathrm{v} \geqslant 0,$$

so that the entropy is nondecreasing. Since equation (4.36) is valid for v or any part of v, and the integrand is assumed continuous, it follows, with the use of the divergence theorem, that

$$\rho\Theta\dot{s} - \rho r + \frac{\partial q_i}{\partial x_i} - \frac{q_i}{\Theta}\frac{\partial\Theta}{\partial x_i} \geqslant 0,$$
$$\rho\Theta\dot{s} - \rho r + \mathrm{div}\boldsymbol{q} - \frac{\boldsymbol{q}}{\Theta}.\nabla\Theta \geqslant 0,$$

and if we substitute for r from equation (4.31) or subtract equation (4.31) we obtain

$$\begin{aligned} &\rho(\Theta\dot{s} - \dot{u}) + \sigma_{ij}d_{ij} - \frac{q_i}{\Theta}\frac{\partial\Theta}{\partial x_i} \geqslant 0, \\ &\rho(\Theta\dot{s} - \dot{u}) + \mathrm{tr}(\boldsymbol{\sigma}\boldsymbol{D}) - \frac{\boldsymbol{q}}{\Theta}.\nabla\Theta \geqslant 0. \end{aligned} \tag{4.37}$$

Equation (4.37) is the local form of the dissipation inequality, or Clausius Duhem inequality, referred to the spatial configuration. As an exercise show that the local form of the dissipation inequality referred to the reference configuration is

$$\begin{aligned} &\rho_0(\widehat{\Theta}\dot{\widehat{s}} - \dot{\widehat{u}}) + \Sigma_{Ki}\frac{\partial v_i}{\partial X_K} - \frac{Q_K}{\widehat{\Theta}}\frac{\partial\widehat{\Theta}}{\partial X_K} \geqslant 0, \\ &\rho_0(\widehat{\Theta}\dot{\widehat{s}} - \dot{\widehat{u}}) + \mathrm{tr}(\Sigma\dot{\boldsymbol{F}}) - \frac{\boldsymbol{Q}}{\widehat{\Theta}}\nabla_X\widehat{\Theta} \geqslant 0, \end{aligned} \tag{4.38}$$

where $\widehat{s}(\boldsymbol{X}, t) = s(\boldsymbol{x}(\boldsymbol{X}, t), t)$, $\widehat{\Theta}(\boldsymbol{X}, t) = \Theta(\boldsymbol{x}(\boldsymbol{X}, t), t)$. In equation (4.38) $\Sigma\dot{\boldsymbol{F}}$ can be replaced by $\boldsymbol{S}\dot{\boldsymbol{E}}$ or $\boldsymbol{S}\dot{\boldsymbol{C}}/2$.

It is often convenient to introduce the specific Helmholtz free energy function, $f = u - Ts$, and substituting for u in equation (4.37) gives

$$
\begin{aligned}
&-\rho(s\dot{\Theta} + \dot{f}) + \sigma_{ij}d_{ij} - \frac{q_i}{\Theta}\frac{\partial\Theta}{\partial x_i} \geqslant 0, \\
&-\rho(s\dot{\Theta} + \dot{f}) + \mathrm{tr}(\boldsymbol{\sigma}\boldsymbol{D}) - \frac{\boldsymbol{q}}{\Theta}.\nabla\Theta \geqslant 0.
\end{aligned}
\tag{4.39}
$$

The equality signs in equations (4.37) to (4.39) hold only for an ideal thermodynamically reversible process. The left-hand sides of equations (4.37) and (4.38) are equal to $\rho_0\Theta\dot{\gamma}$ and $\rho\Theta\dot{\gamma}$, respectively, where $\dot{\gamma}\geqslant 0$ is the rate of entropy production per unit mass. It is interesting to note that the rate of entropy production per unit mass due to heat conduction alone is given by

$$
\dot{\gamma}_{(hc)} = -\frac{q_i}{\rho\Theta^2}\frac{\partial\Theta}{\partial x_i} = -\frac{q}{\rho\Theta^2}.\nabla\Theta, \tag{4.40}
$$

referred to the spatial configuration and

$$
\dot{\gamma}_{hc} - \frac{Q_K}{\rho_0\widehat{\Theta}^2}\frac{\partial\widehat{\Theta}}{\partial X_K} = -\frac{Q}{\rho_0\widehat{\Theta}^2}\cdot\nabla_X\widehat{\Theta}, \tag{4.41}
$$

referred to the reference configuration. In the absence of shock waves, heat conduction is the only source of entropy production in a perfectly elastic solid. This is discussed in later chapters.

EXERCISES

4.1. Obtain the spatial form of the stress power in terms of the deviatoric and isotropic parts of the Cauchy stress tensor $\boldsymbol{\sigma}$ and the rate of deformation tensor $\boldsymbol{D}$.

4.2. Show that for an isotropic linear elastic solid the specific stress power is equal to the rate of change of specific strain energy.

4.3. Obtain the following virtual relations,

$$
\delta\boldsymbol{F} = (\nabla\otimes\delta\boldsymbol{u})\boldsymbol{F} \text{ and } \delta\boldsymbol{F}^{-1} = -\boldsymbol{F}^{-1}(\nabla\otimes\delta\boldsymbol{u}).
$$

4.4. Use the results of problem 3 to show that the virtual variation of Almansi's strain tensor, $\boldsymbol{\eta}$, and $\delta\mathbf{e} = \left(\nabla \otimes \delta u + (\nabla \otimes \delta u)^T\right)/2$ are related by

$$\delta e_{ij} = \delta\eta_{ij} + \frac{\partial \delta u_k}{\partial x_i}\eta_{kj} + \eta_{ik}\frac{\partial \delta u_k}{\partial x_j},$$

$$\delta\mathbf{e} = \delta\boldsymbol{\eta} + (\nabla \otimes \delta\boldsymbol{u})^T\boldsymbol{\eta} + \boldsymbol{\eta}(\nabla \otimes \delta\boldsymbol{u}).$$

4.5. Show that the virtual variation of the Lagrangian strain tensor is given by

$$\delta\boldsymbol{E} = \frac{1}{2}\left\{\boldsymbol{F}^T(\nabla_X \otimes \delta\mathbf{u}) + (\nabla_X \otimes \delta\boldsymbol{u})\boldsymbol{F}\right\}.$$

4.6. Express the referential form of the Clausius Duhem inequality in terms of the second Piola-Kirchhoff stress tensor and the right Cauchy-Green strain tensor.

4.7. Justify the Clausius-Planck inequality

$$\rho(\Theta\dot{s} - \dot{u}) + \mathrm{tr}(\boldsymbol{\sigma}\boldsymbol{D}) \geqslant 0$$

and discuss its significance for an isentropic process.

4.8. Discuss the significance, for a reversible isothermal process, of the Clausius-Planck inequality in the equivalent form

$$-\rho(s\dot{\Theta} + \dot{f}) + \mathrm{tr}(\boldsymbol{\sigma}\boldsymbol{D}) \geqslant 0.$$

REFERENCES

1. Malvern, L.E. (1969). Introduction to the Mechanics of a Continuous Medium. Prentice-Hall, Inc.
2. Haupt, P. (2000). Continuum Mechanics and Theory of Materials. Springer-Verlag.
3. Kestin, J.A. (1966). Course in Thermodynamics, Vol. I. Blaisdell.
4. Muller, I., and Ruggiero, T. (1993). Extended Rational Thermodynamics, 2nd Ed. Springer.
5. Chandrasekaraiah, D.S. (1986). Thermoelasticity with Second Sound. App. Mech. Rev. 39, pp. 355–376.

5 Material Models and Constitutive Equations

5.1 Introduction

The response of material bodies to applied forces and heating has not been considered in previous chapters, except that it is assumed that the materials are nonpolar. In order to obtain relations between stress, deformation, and temperature fields, it is necessary to have constitutive equations. These equations give the response of idealized models of real materials, to the application of forces and heating. In this book electromagnetic effects are not considered.

Admissible constitutive equations must not violate the laws of thermodynamics and should satisfy certain additional principles. In this chapter some classical constitutive equations and other relations for models of actual materials are considered, before discussing these additional principles in detail.

As in previous chapters, certain relations are presented in both suffix and symbolic notation; however, relations that involve tensors of higher order than two are given in suffix notation only.

5.2 Rigid Bodies

A rigid motion of a continuum is given by

$$\begin{aligned} x_i(t) &= Q_{iK}(t)X_K + c_i(t), \\ \boldsymbol{x}(t) &= \boldsymbol{Q}(t)\boldsymbol{X} + \boldsymbol{c}(t), \end{aligned} \tag{5.1}$$

where $\boldsymbol{Q}$ is a two-point proper orthogonal second-order tensor and $\boldsymbol{c}$ has the dimension of length. It follows from equation (5.1) that

$$F_{iK} = Q_{iK}, \ \det[\boldsymbol{F}] = 1, \quad \boldsymbol{C} = \boldsymbol{F}^{\mathrm{T}}\boldsymbol{F} = \boldsymbol{I} \text{ and } \boldsymbol{D} = \boldsymbol{0}.$$

A rigid body may be defined as a body for which the distance between any two particles of the body remains invariant, and the motion given by equation (5.1) satisfies this definition. To show this, take any two particles of the body with position vectors $\boldsymbol{X}_1$ and $\boldsymbol{X}_2$ and corresponding position vectors $\boldsymbol{x}_1$ and $\boldsymbol{x}_2$ after a rigid body motion that involves translation and rotation. It then follows from equation (5.1) that

$$\begin{aligned}(\boldsymbol{x}_1 - \boldsymbol{x}_2) \cdot (\boldsymbol{x}_1 - \boldsymbol{x}_2) &= \boldsymbol{Q}(\boldsymbol{X}_1 - \boldsymbol{X}_2) \cdot \boldsymbol{Q}(\boldsymbol{X}_1 - \boldsymbol{X}_2) \\ &= (\boldsymbol{X}_1 - \boldsymbol{X}_2) \cdot (\boldsymbol{X}_1 - \boldsymbol{X}_2),\end{aligned}$$

since $\boldsymbol{Q}\boldsymbol{Q}^T = \boldsymbol{Q}^T\boldsymbol{Q} = \boldsymbol{I}$. Consequently, the distance between any two particles is invariant for a rigid motion. The rigid body model is a useful idealization for certain real materials in applications where any deformation is negligible for the problem considered.

A necessary and sufficient condition for a rigid motion is

$$d_{ij} \equiv 0, \quad \boldsymbol{D} \equiv 0, \tag{5.2}$$

which implies that the stress power $\sigma_{ij}d_{ij}$ is zero. This is Killing's theorem, which has already been discussed in chapter 2. Equation (5.2) is a constitutive equation for a rigid body, and it follows that $d_{ii} = 0$ is a necessary but not sufficient condition for a body to be rigid but is a necessary and sufficient condition for an incompressible fluid. The stresses in a rigid body are undetermined, the response to the application of forces is governed by equation (5.1) and the laws of classical rigid body mechanics. Mechanical and thermal effects are uncoupled. In addition to equation (5.2), there is a thermal equation of state for a rigid body,

$$u(\Theta) = \int_{\Theta_0}^{\Theta} c\left(\hat{\Theta}\right) d\hat{\Theta}, \tag{5.3}$$

where u, as before, is the specific internal energy, Θ is the absolute temperature, Θ_0 is a reference temperature, and $c(\Theta)$ is the specific heat. The energy equation for a rigid body follows from equation (4.12) and is

$$\begin{aligned}\rho\dot{u} - \rho r + \frac{\partial q_i}{\partial x_i} &= 0, \\ \rho\dot{u} - \rho r + \operatorname{div}\boldsymbol{q} &= 0,\end{aligned} \tag{5.4}$$

where $\boldsymbol{q}$ is the heat flux vector and r is rate of heat supply per unit mass, for example, Joule heating due the passage of an electric current. A semi-empirical law, Fourier's Law of heat conduction, given by

$$q_i = -k_{ij}(\Theta)\frac{\partial \Theta}{\partial x_j}, \quad \boldsymbol{q} = -\boldsymbol{k}(\Theta)\mathrm{grad}\Theta \tag{5.5}$$

is now introduced. In equation (5.5) $\boldsymbol{k}$ is the heat conduction tensor, is symmetric and positive or positive semidefinite, and in general is a function of temperature. The positive or positive semidefinite property is consistent with the Clausius statement of the second law of thermodynamics. Substituting equation (5.5) in equation (5.4) gives the spatial form of the energy equation,

$$\begin{aligned} \rho c(\Theta)\frac{\partial \Theta}{\partial t} - \rho r - \frac{\partial}{\partial x_i}(k_{ij}\frac{\partial \Theta}{\partial x_j}) = 0, \\ \rho c(\Theta)\frac{\partial \Theta}{\partial t} - \rho r - \mathrm{div}(\boldsymbol{k}\,\mathrm{grad}\Theta) = 0. \end{aligned} \tag{5.6}$$

For many problems the temperature range is small enough that the temperature dependence of $\boldsymbol{k}$ and c can be neglected. If in addition the rigid body is isotropic, $\boldsymbol{k} = \boldsymbol{I}k$, and equation (5.6) becomes

$$\rho c\frac{\partial \Theta}{\partial t} - \rho r - k\nabla^2\Theta = 0,$$

where

$$\nabla^2 \equiv \mathrm{divgrad} \equiv \partial^2 \big/ \partial x_i \partial x_i.$$

5.3 Ideal Inviscid Fluid

An ideal inviscid fluid has a simple constitutive equation, namely,

$$\sigma_{ij} = -p\delta_{ij}, \quad \boldsymbol{\sigma} = -p\mathbf{I}, \tag{5.7}$$

where p is a hydrostatic pressure. Equation (5.7) arises since an inviscid fluid, which is a mathematical idealization, can sustain no shearing stresses;

consequently, the Cauchy stress must be a pure hydrostatic stress with no deviatoric part. A thermal equation of state is required in addition to equation (5.7), for example, if the fluid is a perfect gas, that is, a gas that obeys Boyle's law and Charles law, then

$$\frac{p}{\rho} = R\Theta, \tag{5.8}$$

where R is the gas constant for the particular gas, Θ is the absolute temperature, p is the absolute pressure, and ρ is the density. Since the internal energy of a perfect gas can be expressed as a function of temperature only, equation (5.3) also holds for a perfect gas if $c(\Theta)$ is replaced by $c_v(\Theta)$, the specific heat at constant volume. The stress power of a compressible inviscid fluid is $P = -p\dot{v}$ where $v = 1/\rho$.

If we consider a viscous fluid, equation (5.7) holds only for equilibrium states and is not a constitutive relation.

5.4 Incompressible Inviscid Fluid

This is an inviscid fluid model that can only undergo isochoric flow. The constitutive equation chosen to model a given material may be influenced by the problem considered. For example, for some problems involving a liquid, such as gravity water waves, it may be realistic to assume an incompressible inviscid fluid, whereas for other problems, such as water hammer in a pipe, compressibility effects must be considered, and for boundary layer problems, viscous effects are dominant.

If a fluid is assumed to be mechanically incompressible the coefficient of volume thermal expansion must be zero so that only isochoric deformation is possible. It has been shown by Müller [1] that a fluid model that is mechanically incompressible cannot have a nonzero coefficient of volume thermal expansion, otherwise the second law of thermodynamics is violated. Since an incompressible fluid can undergo only isochoric deformation,

$$d_{ii} = 0, \quad \mathrm{tr}\boldsymbol{D} = 0.$$

The stress power for an incompressible inviscid fluid is zero and equation (5.3) holds.

5.5 Newtonian Viscous Fluid

First we consider the compressible case. The Newtonian viscous fluid model has the following properties. (1) Stress components $\sigma_{ij} + p\delta_{ij}$, where p is the thermodynamic pressure, are linear homogeneous functions of d_{ij}. The thermodynamic pressure is the pressure that appears in the thermodynamic equation of state, for example, for a perfect gas p is given by equation (5.8). (2) The fluid is isotropic. (3) When the rate of deformation is zero, the stress is hydrostatic, that is, $\sigma_{ij} = -p\delta_{ij}$. It follows from (1) to (3) that $\sigma_{ij} + p\delta_{ij} = C_{ijkl}d_{kl}$, where C_{ijkl} are the components of a fourth-order isotropic tensor. Since the most general fourth-order isotropic tensor has components given by

$$C_{ijkl} = \alpha\delta_{ij}\delta_{kl} + \beta\delta_{ik}\delta_{jl} + \gamma\delta_{il}\delta_{jk},$$

where α, β, and γ are scalars, $\sigma_{ij} + p\delta_{ij} = \left(\alpha\delta_{ij}\delta_{kl} + \beta\delta_{ik}\delta_{jl} + \gamma\delta_{il}\delta_{jk}\right)d_{kl}$, for a single phase fluid. Since $\boldsymbol{\sigma}$ and $\boldsymbol{D}$ are symmetric, this can be put in the form

$$\begin{aligned} \sigma_{ij} + p\delta_{ij} &= 2\eta d_{ij} + \varsigma d_{kk}\delta_{ij} \\ \boldsymbol{\sigma} + p\boldsymbol{I} &= 2\eta\boldsymbol{D} + \varsigma\boldsymbol{I}\mathrm{tr}\boldsymbol{D}, \end{aligned} \tag{5.9}$$

where η and ς are viscosity coefficients. For isochoric flow it follows from the constitutive equation (5.9) that

$$\sigma_{ij} + p\delta_{ij} = 2\eta d_{ij}, \quad d_{ii} = 0,$$

which indicates that η is the viscosity component for simple shear. For pure dilatation it follows from equation (5.9) that

$$\frac{\sigma_{kk}}{3} + p = \xi d_{kk}, \tag{5.10}$$

where $\xi = 2\eta/3 + \varsigma$ is the bulk viscosity. Substitution of equation (5.9) in the equation of motion (3.15),

$$\frac{\partial\sigma_{ij}}{\partial x_j} + \rho f_i = \rho\frac{dv_i}{dt},$$

with the use of equation (5.10) gives

$$-\frac{\partial p}{\partial x_i}+\eta\frac{\partial^2 v_i}{\partial x_j \partial x_j}+\left(\xi+\frac{1}{3}\eta\right)\frac{\partial^2 v_j}{\partial x_i \partial x_j}+\rho f_i=\rho\frac{dv_i}{dt}. \tag{5.11}$$

This equation is the Navier-Stokes equation and is nonlinear since $dv_i/dt = \partial v_i/\partial t + v_j \partial v_i/\partial x_j$ is nonlinear. It should be noted that the mechanical pressure $-\sigma_{kk}/3$ is not equal to the thermodynamic pressure p unless it is assumed that $\xi = 0$. This assumption is known as Stokes' hypothesis and is a good approximation for monatomic gases at relatively low pressures. If $\xi = 0$, it may be deduced from equation (5.10) that there is no dissipation of mechanical work due to viscosity when the fluid undergoes dilatation with no distortion. According to White [2], the bulk viscosity property has been a controversial concept.

The viscosity coefficients η and ξ are transport properties and are, in general, functions of temperature. For a liquid η decreases and for a gas η increases as the temperature increases.

An incompressible Newtonian fluid is a useful model for many liquids. This model has only one viscosity coefficient η and equation (5.9) is replaced by

$$\sigma_{ij} = 2\eta d_{ij} - p\delta_{ij}, \tag{5.12}$$

with $d_{ii} = 0$, where $p = -\sigma_{kk}/3$ is a Lagrangian multiplier that is obtained from stress boundary conditions. A more extensive treatment of a Newtonian viscous fluid is given in [2] and some other aspects are given in chapter 9.

5.6 Classical Elasticity

An elastic solid is a material for which the stress depends only on the deformation, when the deformation is isothermal. Classical elasticity is a linear elastic theory based on the assumption that the displacement gradient tensor is infinitesimal so that the strain and rotation tensors are linearized and the distinction between the spatial and referential forms of the governing equations disappears. There is a vast literature on classical elasticity dating from the mid-nineteenth century onward and only a very brief discussion of the stress strain relations is given here. The book by Love [3] is an extensive exposition of linear elasticity and contains references up to 1926. Two significant later general treatises on the subject are Goodier and Timoshenko [4] and Sokolnikov [5], and there are many books that consider specialized areas. Here we briefly discuss some basic features of the constitutive relations of classical elasticity for isothermal deformation. The components of Cauchy stress, σ_{ij}, are linear functions of the components of the infinitesimal strain tensor

$$e_{ij} = (1/2)\big(\partial u_i/\partial x_j + \partial u_j/\partial x_i\big).$$

Since σ_{ij} and e_{ij} are the components of second-order symmetric tensors,

$$\sigma_{ij} = c_{ijkl} e_{kl}, \tag{5.13}$$

where the c_{ijkl} are the components of a fourth-order tensor known as the stiffness tensor, with symmetries.

$$c_{ijkl} = c_{jikl} = c_{jilk},$$

since $\sigma_{ij} = \sigma_{ji}$ and $e_{kl} = e_{lk}$. Equation (5.13) is sometimes known as the generalized Hooke's law. It can be shown that there is a further symmetry, $c_{ijkl} = c_{klij}$ if a strain energy function,

$$w(\boldsymbol{e}) = \frac{1}{2} c_{ijkl} e_{ij} e_{kl}, \tag{5.14}$$

exists, such that

$$\sigma_{ij} = \frac{\partial w(\boldsymbol{e})}{\partial e_{ij}}. \tag{5.15}$$

It then follows that a classical linear elastic solid can have at most twenty-one independent elastic constants. It should be noted that, in equation

(5.14), the symmetry of the infinitesimal strain tensor $\boldsymbol{e}$ is ignored for the differentiation (5.15). Sokolnikov [5] has given more extensive discussion of the symmetries of the stiffness tensor. Equation (5.13) can be inverted to give

$$e_{ij} = a_{ijkl}\sigma_{kl}, \tag{5.16}$$

where a_{ijkl} are the components of a fourth-order tensor known as the compliance tensor, which has the same symmetries as the stiffness tensor. The strain energy function in terms of the stress components, sometimes known as the complementary energy per unit volume, is

$$w_c = \frac{1}{2}a_{ijkl}\sigma_{ij}\sigma_{kl},$$

and equation (5.16) can be obtained from

$$e_{ij} = \frac{\partial w_c(\boldsymbol{\sigma})}{\partial \sigma_{ij}}.$$

We will consider a homogeneous solid so that the components of the stiffness and compliant tensors are constants.

As already mentioned the components of the most general isotropic fourth-order tensor are of the form

$$\alpha\delta_{ij}\delta_{kl} + \beta\delta_{ik}\delta_{jl} + \gamma\delta_{il}\delta_{jk},$$

where α, β, and γ are scalars. It then follows that for an isotropic linear elastic solid, equation (5.13) is of the form,

$$\sigma_{ij} = \alpha\delta_{ij}e_{kk} + \beta e_{ij} + \gamma e_{ji}.$$

Since $e_{ij} = e_{ji}$ it then follows that

$$c_{ijkl} = 2\mu\delta_{ik}\delta_{jl} + \lambda\delta_{ij}\delta_{kl}, \tag{5.17}$$

and from equation (5.13) that

$$\sigma_{ij} = 2\mu e_{ij} + \lambda e_{kk}\delta_{ij}, \tag{5.18}$$

where μ and λ are known as Lamé's constants for isothermal deformation. A compressible classical linear elastic solid has two independent elastic constants, and any other elastic constant can be expressed in terms of μ and λ. For example, Young's modulus, the bulk modulus, and Poisson's ratio are given by

$$E = \frac{\mu(3\lambda + 2\mu)}{\lambda + \mu}, \quad K = \lambda + \frac{2}{3}\mu, \quad \text{and} \quad \nu = \frac{\lambda}{2(\lambda + \mu)},$$

respectively. Equation (5.18) can be inverted to give

$$e_{ij} = \frac{1}{2\mu}\sigma_{ij} + \frac{-\lambda\delta_{ij}}{2\mu(3\lambda + 2\mu)}\sigma_{kk}. \tag{5.19}$$

A brief discussion of some aspects of anisotropic linear elasticity is now given. Equation (5.13) can be expressed in the matrix form

$$[\sigma] = [C][e], \tag{5.20}$$

where $[\sigma]$ and $[e]$ are (6×1) column matrices related to the Cauchy stress and linear infinitesimal strain tensors by

$$[\sigma]^T = [\sigma_1\ \sigma_2\ \sigma_3\ \sigma_4\ \sigma_5\ \sigma_6] = [\sigma_{11}\ \sigma_{22}\ \sigma_{33}\ \sigma_{23}\ \sigma_{31}\ \sigma_{12}],$$
$$[e]^T = [e_1\ e_2\ e_3\ e_4\ e_5\ e_6] = [e_{11}\ e_{22}\ e_{33}\ 2e_{23}\ 2e_{31}\ 2e_{12}],$$

and $[C]$ is a (6×6) symmetric matrix related to the fourth-order stiffness tensor by

$$[C] = \begin{bmatrix} C_{11}=c_{1111} & C_{12}=c_{1122} & C_{13}=c_{1133} & C_{14}=c_{1123} & C_{15}=c_{1113} & C_{16}=c_{1112} \\ & C_{22}=c_{2222} & C_{23}=c_{2233} & C_{24}=c_{2223} & C_{25}=c_{2231} & C_{26}=c_{2212} \\ & & C_{33}=c_{3333} & C_{34}=c_{3323} & C_{35}=c_{3331} & C_{36}=c_{3312} \\ & \text{Symmetric} & & C_{44}=\frac{c_{2323}+c_{2332}}{2} & C_{45}=c_{2331} & C_{46}=c_{2312} \\ & & & & C_{55}=\frac{c_{3131}+c_{3113}}{2} & C_{56}=c_{3112} \\ & & & & & C_{66}=\frac{c_{1212}+c_{1221}}{2} \end{bmatrix}. \tag{5.21}$$

The matrix $[C]$ is known as the Voigt matrix. It follows from equation (5.21) that for an isotropic solid,

$$[C] = \begin{bmatrix} 2\mu+\lambda & \lambda & \lambda & 0 & 0 & 0 \\ & 2\mu+\lambda & \lambda & 0 & 0 & 0 \\ & & 2\mu+\lambda & 0 & 0 & 0 \\ & \text{Symmetric} & & \mu & 0 & 0 \\ & & & & \mu & 0 \\ & & & & & \mu \end{bmatrix}.$$

Equation (5.20) can be expressed in suffix notation,

$$\sigma_p = \sum_{q=1}^{6} C_{pq} e_q, \quad p \in \{1, 2. \ldots 6\};$$

however, it is important to note that this is a matrix equation, not a tensor equation. The compliance relation (5.16) can also be expressed in matrix form as

$$[e] = [A][\sigma],$$

where $[A]$ is the matrix inverse of $[C]$.

5.7 Linear Thermoelasticity

In this section some constitutive aspects of linear thermoelasticity are considered. These are restricted to isotropic solids and the assumptions of isothermal linear elasticity, along with small deviations from a reference absolute temperature Θ_o, so that $|\Theta - \Theta_0|/\Theta_0 << 1$. When the temperature of an isotropic unstressed elastic solid is changed, pure dilatation $e = \alpha_v(\Theta - \Theta_o)$ occurs so that the components of the thermal strain e^t_{ij} are

$$e^t_{ij} = \frac{\alpha_v}{3}(\Theta - \Theta_0)\delta_{ij}, \tag{5.22}$$

where α_v is the volume coefficient of thermal expansion, which is assumed to be constant.

It is clear from equation (5.22) that no shear deformation is produced by a uniform temperature change in a homogeneous elastic solid. For homogeneous deformation with strain e_{ij}, the part due to the stress is $e_{ij} - e^t_{ij}$, and it follows from superposition and equations (5.18) and (5.22) that

$$\sigma_{ij} = 2\mu e_{ij} + \lambda e_{kk}\delta_{ij} - \left(\frac{2}{3}\mu + \lambda\right)\alpha_v(\Theta - \Theta_0)\delta_{ij}, \tag{5.23}$$

which is an equation of state. In equation (5.23) μ and λ are the isothermal moduli, assumed constant for infinitesimal deformation and $|\Theta - \Theta_0|/\Theta_0 << 1$. In this section specific thermodynamic properties are taken per unit mass. The stress and infinitesimal strain $\boldsymbol{\sigma}$ and $\boldsymbol{e}$ are now decomposed into the deviatoric and isotropic parts to illustrate that thermodynamic effects in an isotropic linear compressible thermoelastic solid are confined to the dilatational part of the deformation. Substitution of $\sigma_{ij} = s_{ij} + \sigma\delta_{ij}$ and $e_{ij} = e'_{ij} + \frac{e}{3}\delta_{ij}$, where $e = e_{ii}$, $\sigma = \sigma_{ii}/3$, in equation (5.23) gives the decoupled shear and dilatational responses,

$$s_{ij} = 2\mu e'_{ij}, \quad \sigma = K\{e - \alpha_v(\Theta - \Theta_0)\}, \tag{5.24}$$

where $K = 2\mu/3 + \lambda$ is the isothermal bulk modulus.

It is evident from equation (5.24) that the uncoupled shear deformation is not influenced by temperature change and is both isothermal and isentropic, that is, for a linear thermoelastic solid the thermodynamic effects are restricted to the dilatational part of the deformation.

There are four important specific thermoelastic potentials each of which can be expressed as a function of two of the thermodynamic properties $\boldsymbol{\sigma}$, $\boldsymbol{e}$, s, and Θ. A potential, which is expressed as a function of two of the above thermodynanic properties, such that the other two can be obtained by partial differentiation is a fundamental equation of state. These fundamental equations of state are related by Legendre transformations.

First the specific internal energy and Helmholtz free energy are considered and are expressed as fundamental equations of state in the form

$$u = \widehat{u}\left(e'_{ij}, e, s\right) \tag{5.25}$$

and

$$f = \hat{f}(e'_{ij}, e, \Theta), \tag{5.26}$$

respectively, where $\hat{u}(0, 0, s_0) = 0$ and $f(0, 0, \Theta_0) = 0$. The strain tenor $\boldsymbol{e}$ is replaced by $\boldsymbol{e}' + (e/3)\boldsymbol{I}$ in equations (5.25) and (5.26) in order to indicate the decoupling of the distortional and dilatational effects.

A Gibbs relation for a linear isotropic thermoelastic solid is

$$\Theta ds = du - \frac{\sigma_{ij} de_{ij}}{\rho} = du - \frac{s_{ij} de'_{ij} - \sigma de}{\rho}. \tag{5.27}$$

It follows from equation (5.27) and

$$du = \frac{\partial \hat{u}}{\partial e'_{ij}} de'_{ij} + \frac{\partial \hat{u}}{\partial e} de + \frac{\partial \hat{u}}{\partial s} ds$$

that

$$s_{ij} = \rho \frac{\partial \hat{u}}{\partial e'_{ij}}, \quad \sigma = \rho \frac{\partial \hat{u}}{\partial e}, \quad \Theta = \rho \frac{\partial \hat{u}}{\partial s}. \tag{5.28}$$

Equation (5.28) indicates that the fundamental equation of state for the internal energy is of the form $u = \hat{u}(e'_{ij}, e, s)$.

A Gibbs relation

$$df = -sd\Theta + \frac{\sigma_{ij} de_{ij}}{\rho} = -sd\Theta + \frac{s_{ij} de'_{ij} + \sigma de}{\rho} \tag{5.29}$$

is obtained from equation (5.27) and the Legendre transformation $f = u - \Theta s$, and it follows from equation (5.29) and

$$df = \frac{\partial f}{\partial e'_{ij}} de_{ij} + \frac{\partial f}{\partial e} de + \frac{\partial f}{\partial \Theta} d\Theta$$

that

$$s_{ij} = \rho \frac{\partial \widehat{f}}{\partial e'_{ij}}, \quad \sigma = \rho \frac{\partial \widehat{f}}{\partial e}, \quad s = -\frac{\partial \widehat{f}}{\partial \Theta}. \tag{5.30}$$

Equations (5.30) indicate that the fundamental equation of state for the Helmholtz free energy is of the form $f = \widehat{f}(e'_{ij}, e, s)$.

Substituting equation (5.24) in equation (5.29) and integrating gives

$$\widehat{f}(e'_{ij}, e, \Theta) = \frac{1}{\rho}\left(\mu e'_{ij} e'_{ij} + \frac{1}{2} K e^2 - K\alpha_v e(\Theta - \Theta_0)\right) + F(\Theta), \tag{5.31}$$

where $F(\Theta)$ is a function to be determined. Equation (5.30) is consistent with equations (5.24) and $(5.30)_{1,2}$. It follows from $(5.30)_3$ and (5.31) that

$$s = \frac{K\alpha_v e}{\rho} - F'(\Theta). \tag{5.32}$$

Since $(\Theta ds)_{e=0} = c_e d\Theta$, where c_e is the specific heat at constant strain $\boldsymbol{e}$, and $s_{e=0} = -F'(\theta)$,

$$-F'(\Theta) = c_e \int_{\Theta_0}^{\Theta} \frac{d\widehat{\Theta}}{\widehat{\Theta}} = \ln \frac{\theta}{\theta_0}. \tag{5.33}$$

The linearized form of equation (5.33) is

$$-F'(\Theta) = c_e \left(\frac{\Theta - \Theta_0}{\Theta_0}\right), \tag{5.34}$$

and this along with equation (5.32) gives

$$s = \frac{K\alpha_v e}{\rho} + c_e \left(\frac{\Theta - \Theta_0}{\Theta_0}\right). \tag{5.35}$$

According to equation (5.35) the specific entropy is taken as zero for a reference state with $\Theta = \Theta_0$ and $e = 0$.

Substituting the integrated form of equation (5.34) in equation (5.31) gives the fundamental equation of state

$$\hat{f}(e'_{ij}, e, \Theta) = \frac{1}{\rho}\left(\mu e'_{ij}e'_{ij} + \frac{1}{2}Ke^2 - K\alpha_v e(\Theta - \Theta_0)\right) - \frac{1}{2}c_e\frac{(\Theta - \Theta_0)^2}{\Theta_0}. \tag{5.36}$$

It may be verified that equation (5.30) can be obtained from equation (5.36). The strain energy per unit volume for isothermal deformation,

$$w(\mathbf{e}', e) = \rho f(\mathbf{e}', e, \Theta_0) = \mu e'_{ij}e'_{ij} + \frac{1}{2}Ke^2,$$

is obtained from equation (5.36).

It follows from equation (5.35) that, for isentropic deformation from the reference state,

$$\Theta - \Theta_0 = -\frac{K\alpha_v e\Theta_0}{\rho c_v}$$

and substitution in equation $(5.24)_2$ gives

$$\sigma = \left(K + \frac{\alpha_v^2 K^2 \Theta_0}{\rho c_e}\right)e.$$

Consequently, the isentropic (reversible adiabatic) bulk modulus K_a is given by

$$K_a = K + \frac{\alpha_v^2 K^2 \Theta_0}{\rho c_e},$$

and the isothermal and isentropic shear moduli are equal. It can be deduced from the Gibbs relation (5.27) that the following expression for the specific internal energy

$$\hat{u}(e'_{ij}, e, \Theta) = \frac{1}{\rho}\left(\mu e'_{ij} e'_{ij} + \frac{1}{2}Ke^2 + K\alpha_v e\Theta_0\right) + \frac{1}{2}c_e\frac{(\Theta - \Theta_0)^2}{\Theta_0} + c_e(\Theta - \Theta_0) \tag{5.37}$$

is obtained from the Legendre transformation $u = f + \Theta s$ and equations (5.35) and (5.36). It should be noted that equation (5.37) is not a fundamental equation of state since $\hat{u}$ is a function of e'_{ij}, e and Θ instead of e'_{ij}, e and s. It is easily determined from equation (5.37) that

$$c_e = \left(\frac{\partial \hat{u}}{\partial \Theta}\right)_{\Theta=\Theta_0}.$$

The fundamental equation of state

$$\tilde{u}(e'_{ij}, e, s) = \frac{1}{\rho}\left(\mu e'_{ij} e'_{ij} + \frac{1}{2}Ke^2 + K\alpha_v e\Theta_0\right) + c_e\left\{\left(\frac{s}{c_e}\Theta_0 - \frac{K\alpha_v e\Theta_o}{\rho c_e}\right) + \frac{1}{2\Theta_0}\left(\frac{s}{c_e}\Theta_0 - \frac{K\alpha_v e\Theta_o}{\rho c_e}\right)^2\right\} \tag{5.38}$$

is obtained by eliminating $(\Theta - \Theta_0)$ from equations (5.35) and (5.37). It may be verified that equations (5.28) can be obtained from equation (5.38) with the use of equation (5.35).

The specific enthalpy for a linear thermoelastic solid is related to the internal energy by the Legendre transformation $h = u - s_{ij}e'_{ij}/\rho - \sigma e/\rho$, and the differential of this along with equation (5.27) gives the Gibbs relation

$$dh = \Theta ds - \left(\frac{e'_{ij}\, ds_{ij} + ed\sigma}{\rho}\right). \tag{5.39}$$

It follows from equation (5.39) that

$$dh = \frac{\partial h}{\partial s_{ij}} ds_{ij} + \frac{\partial h}{\partial \sigma} d\sigma + \frac{\partial h}{\partial s} ds$$

and

$$e'_{ij} = \rho \frac{\partial h}{\partial s_{ij}}, \quad e = \rho \frac{\partial h}{\partial \sigma}, \quad \Theta = \frac{\partial h}{\partial s}. \tag{5.40}$$

Equation (5.40) indicates that the fundamental equation of state for the enthalpy is of the form $h = h(s_{ij}, \sigma, s)$. Before determining the fundamental form $h(s_{ij}, \sigma, s)$ it useful to consider the form

$$\breve{h}(s_{ij},\sigma,\Theta) = \frac{s_{ij}s_{ij}}{2\rho\mu} + \frac{1}{2}\frac{K}{\rho}\left(\frac{\sigma}{K} + \alpha_v(\Theta - \Theta_0)\right)^2 +$$
$$\left(\frac{K\alpha_v\Theta_0}{\rho} - \frac{\sigma}{\rho}\right)\left(\frac{\sigma}{K} + \alpha_v(\Theta - \Theta_0)\right) + c_e(\Theta - \Theta_0) + \frac{1}{2}c_e\frac{(\Theta - \Theta_0)^2}{\Theta_0}, \tag{5.41}$$

that is determined from equations (5.24), (5.38), and $h = u - s_{ij}e'_{ij}/\rho - \sigma e/\rho$. It may be deduced from equation (5.39) that, for a reversible process at constant stress, the specific heart at constant stress is given by

$$c_\sigma = \left(\frac{\partial \breve{h}}{\partial \Theta}\right)_{\Theta = \Theta_0}.$$

It then follows from equation (5.41) that

$$c_\sigma = c_e + \frac{K\alpha_v^2\Theta_0}{\rho}.$$

A fundamental equation of state for the enthalpy is obtained by eliminating $(\Theta_0 - \Theta)$ from equation (5.41) and

$$\Theta - \Theta_0 = \left(s - \frac{\alpha_v\sigma}{\rho}\right) \bigg/ \left(\frac{K\alpha_v^2}{\rho} + \frac{c_e}{\theta_0}\right),$$

that is obtained from equations $(5.24)_2$ and (5.35).

The Gibbs free energy is the final thermodynamic potential for a linear thermoelastic solid that we will consider. It is related to the Helmholtz free energy by the Legendre transformation, $g = f - s_{ij}s_{ij}/\rho - \sigma e/\rho$, which gives

$$dg = -sd\Theta - \frac{e'_{ij}ds_{ij} - ed\sigma}{\rho}. \tag{5.42}$$

It follows from equation (5.42) and

$$dg = \frac{\partial g}{\partial s_{ij}}ds_{ij} + \frac{\partial g}{\partial \sigma}d\sigma + \frac{\partial g}{\partial \Theta}d\Theta$$

that

$$e'_{ij} = -\rho\frac{\partial g}{\partial s_{ij}}, \quad e = -\rho\frac{\partial g}{\partial \sigma}, \quad s = -\frac{\partial g}{\partial \Theta}. \tag{5.43}$$

Equation (5.43) indicates that the fundamental equation of state for the Gibbs free energy is of the form $g = g(s_{ij}, \sigma, \Theta)$, which can be obtained as

$$g(s_{ij},\sigma,\Theta) = \frac{1}{\rho}\left\{-\frac{s_{ij}s_{ij}}{4\mu} + \frac{K}{2}\left[\frac{\sigma}{K} + \alpha_v(\Theta - \Theta_0)\right]^2 - \left[\frac{\sigma}{K} + \alpha_v(\Theta - \Theta_0)\right]\right.$$
$$\left.[K\alpha_v(\Theta - \Theta_0) + \sigma]\right\} - \frac{1}{2}c_e\frac{(\Theta - \Theta_0)^2}{\Theta_0}, \tag{5.44}$$

by using the Legendre transformation $g = f - (s_{ij}e'_{ij} + \sigma e)/\rho$, and equations (5.24) and (5.36).

As an exercise the reader should verify that equation (5.43) is satisfied by equation (5.44).

The complementary energy per unit volume for isothermal deformation is given by $w_c(\boldsymbol{s}, \sigma) = \rho g(\boldsymbol{s}, \sigma, \theta_0)$, and it follows from equation (5.44) that

$$w_s(s_{ij}, \sigma) = -\frac{s_{ij}s_{ij}}{4\mu} - \frac{\sigma^2}{2K}, \tag{5.45}$$

which is equal to the negative of $w(\boldsymbol{e}', e)$, the strain energy per unit volume for isothermal deformation. Often the complementary energy per unit volume is described as the strain energy in terms of stress with the negative signs in equation (5.45) replaced by positive signs. However, the form (5.45) with negative signs is obtained from the Legendre transformation, $g = f - \left(s_{ij}e'_{ij} + \sigma e\right)/\rho$.

Constitutive relations for nonlinear elastic solids are considered in later chapters.

5.8 Determinism, Local Action, and Material Frame Indifference

It has already been noted that constitutive equations should not violate the laws of thermodynamics. In addition they should satisfy the principles of determinism, local action, and frame indifference. These principles are sometimes called axioms, and it is possible to formulate more than three. For example, Eringen [6] lists eight axioms of constitutive theory; however, the three principles discussed here appear to be sufficient to develop a constitutive theory.

The principle of determinism states that the stress and specific internal energy at a material point at time t are determined uniquely by the past history of the motion and temperature field up to and including time t. The heat flux $\boldsymbol{q}$ is also so determined; however, this is usually of less importance in our considerations than the stress. It is reasonable to assume that the history of the motion and temperature fields in the recent past has more influence on the state of the medium than that of the distant past, that is, the material has a fading memory. A perfectly elastic solid has a perfect memory of an undeformed reference configuration and an inviscid fluid or a Newtonian viscous fluid has no memory of a reference configuration.

The principle of local action states that the stress, specific internal energy, and heat flux at a material point $\boldsymbol{X}$ are independent of the history of motion and temperature outside an arbitrary small neighborhood of $\boldsymbol{X}$. Materials that obey this principle are known as simple materials. In a simple material, the stress at time t at a material point depends only on the history of the deformation gradient and temperature. This implies that all relations between stress and past history of motion can be obtained from experiments that involve homogeneous deformation and temperature. These relations are constitutive equations. In this book we consider only simple materials, since all the usual material models considered in civil and mechanical engineering refer to simple materials.

Before discussing the principle of frame indifference it is necessary to introduce the concepts of a frame of reference and an observer that are often regarded as synonymous. A frame of reference may be regarded as a hypothetical rigid body to which an infinite number of coordinate systems may be attached. Right-handed, rectangular Cartesian coordinate systems are considered for the following discussion.

The concept of a frame of reference involves an observer that can measure time and the relative position of a point in E_3. Two different observers are said to be equivalent if they observe the same distances between pairs of points in E_3 and the same interval of time between events.

Transformation relations between coordinate systems attached to the same frame of reference are time independent. However, transformation relations between coordinate systems attached to different frames of reference are, in general, time dependent unless the frames are moving in relative translation. Equations or physical quantities, which are invariant under coordinate transformation in the same frame of reference, must be tensors or tensor equations; however, this does not necessarily ensure invariance for transformations between coordinate systems attached to different frames of reference. Physical quantities are said to be objective if observers on different frames, in relative motion, observe the same physical quantities, that is, if the quantities are invariant under all changes of frame. Temperature is an example of an objective scalar, velocity is an example of a vector that is, intuitively, not objective. Applied to constitutive equations the principle of material frame indifference, sometimes known as the principle of material objectivity, requires that equivalent observers observe the same material properties or constitutive equations. An alternative statement is, constitutive equations are frame indifferent if they are invariant under all changes of frame. A simple example is the extension of an elastic spring due to an axial force. Equivalent observers observe the same extension of the spring and, since the axial force is related to the extension, the various observers observe the same force. We can assert that contact forces are objective; however, body forces may not be. For example, an observer attached to a body in free fall, under the action of the earth's gravitational field, observes no body force since no acceleration is observed. However, an observer fixed in space observes a body force. A tensor is said to be objective if equivalent observers on different frames of reference observe the same tensor, so that the tensor is invariant under change of frame. The precise meaning of this statement is discussed later in this chapter.

Consider the ordered pair $\{\boldsymbol{x}, t\}$, where $\boldsymbol{x}$ is the position vector of a point in E_3. A change of frame or observer is a one-to-one mapping of $E_3 \times t$ onto itself so that the same distance between any two points in E_3 is observed and time intervals between events are preserved. Let $\boldsymbol{x}$ and t be the position vector, of a point P in E_3, relative to an origin O, and time, respectively, in frame of reference F, and let $\boldsymbol{x}^*$ and $t^* = t - a$, where a is a constant, be the position vector of P relative to an origin O^*, and time, respectively, in frame of reference F*. It should be noted that $\boldsymbol{x}$ is observed by observer F and $\boldsymbol{x}^*$ by observer F*. If F is moving in translation, relative to F*, we have the trivial relation

$$\boldsymbol{x}^* = \boldsymbol{c}(\boldsymbol{t}) + x,$$

where $\boldsymbol{c}(\boldsymbol{t})$ is the position vector of O with respect to O^*. However, if F* and F are undergoing relative translation and rotation,

$$\boldsymbol{x}^* = \boldsymbol{c}(t) + \boldsymbol{Q}(t)\boldsymbol{x}, \quad t = t^* - a, \tag{5.46}$$

where $\boldsymbol{Q}$ is a proper orthogonal tensor, which represents a rotation of F relative to F*, and a is a constant so that two time scales have the same unit and a different origin. Equation (5.46) is known as a Euclidean transformation and is an observer transformation from F to F* since observer F observes position vector $\boldsymbol{x}$ and observer F* observes position vector $\boldsymbol{x}^*$ for the same point in E_3. It may be easier to visualize the Euclidean transformation (5.46) when it is expressed in suffix notation related to coordinate systems $O^*x_i^*$ and Ox_i fixed in F*and F, respectively,

$$x_i^* - c_i(t) = Q_{ij}(t)x_j.$$

When $\boldsymbol{Q} = \boldsymbol{I}$ and $\boldsymbol{c}(t)$ is constant, the transformation is said to be Galilean. A Galilean frame of reference is a frame with respect to which the laws of classical mechanics hold.

If $\boldsymbol{x}_1$ and $\boldsymbol{x}_2$ are position vectors in F of two points it follows from equation (5.46) that

$$\left(\boldsymbol{x}_2^* - \boldsymbol{x}_1^*\right) = \boldsymbol{Q}(\boldsymbol{x}_2 - \boldsymbol{x}_1). \tag{5.47}$$

It is easily verified from equation (5.47) that the condition

$$\left|\boldsymbol{x}_2^* - \boldsymbol{x}_1^*\right| = |\boldsymbol{x}_2 - \boldsymbol{x}_1|,$$

for equivalent observers is satisfied and that is recovered when $\boldsymbol{Q} = \boldsymbol{I}$. It follows from equation (5.47) that a vector $\boldsymbol{u}$ is objective if and only if it satisfies the condition

$$\boldsymbol{u}^*(\boldsymbol{x}^*, t^*) = \boldsymbol{Q}(t)\boldsymbol{u}(x, t), \tag{5.48}$$

for all F and F*, as does the directed line segment $(\boldsymbol{x}_2 - \boldsymbol{x}_1)$, so that different observers observe the same vector. In order to clarify this, let $\{\mathbf{e}_1, \mathbf{e}_2, \mathbf{e}_3\}$ be the unit base vectors for coordinate system $0x_i$, and $\left\{\mathbf{e}_1^*, \mathbf{e}_2^*, \mathbf{e}_3^*\right\}$ the unit base vectors for $0^*x_i^*$, given by

$$\mathbf{e}_i^*(t) = \boldsymbol{Q}(t)\mathbf{e}_i. \tag{5.49}$$

The relation

$$\mathbf{e}_i \otimes \mathbf{e}_i = \boldsymbol{I}$$

and equation (5.49) give

$$\mathbf{e}_i^*(t) \otimes \mathbf{e}_i(t) = \boldsymbol{Q}(t). \tag{5.50}$$

Eliminating Q from equation (5.48) and equation (5.50), gives

$$\boldsymbol{u}^*(\boldsymbol{x}^*, t^*) \cdot \boldsymbol{e}_i^* = \boldsymbol{u}(x, t) \cdot \mathbf{e}_i. \tag{5.51}$$

It may be deduced from equation (5.51) that the components of an objective vector $\boldsymbol{u}$ referred to the coordinate system $0x_i$ are the same as the components of $\boldsymbol{u}^*$ referred to the coordinate system $0^*x_i^*$. Similarly, a tensor $\boldsymbol{T}(\boldsymbol{x}, t)$ referred to F is objective if $\boldsymbol{T}^*(\boldsymbol{x}^*, t^*)$ referred to F*, and $\boldsymbol{T}$ are related by

$$\boldsymbol{T}^* = \boldsymbol{Q}(t)\boldsymbol{T}\boldsymbol{Q}^T(t). \tag{5.52}$$

Substitution of equation (5.50) in equation (5.52) gives

$$\mathbf{e}_i \cdot \boldsymbol{T}\mathbf{e}_j = \mathbf{e}_i^* \cdot \boldsymbol{T}^*\mathbf{e}_j^*. \tag{5.53}$$

It may be deduced from equation (5.53) that the components of an objective second-order tensor $\boldsymbol{T}$ referred to the coordinate system $0x_i$ in F are the same as the components of $\boldsymbol{T}^*$ referred to the coordinate system $0^*\boldsymbol{x}_i^*$ in F*.

It may also be deduced from equation (5.48) that the scalar product of two objective vectors is an objective scalar $\phi^*(\boldsymbol{x}^*, t^*) = \phi(\boldsymbol{x}, t)$ and that a second-order objective tensor $\boldsymbol{T}$ operating on an objective vector $\boldsymbol{u}$ results in an objective vector since

$$\boldsymbol{T}^*\boldsymbol{u}^* = \boldsymbol{Q}\boldsymbol{T}\boldsymbol{Q}^T\boldsymbol{Q}\boldsymbol{u} = \boldsymbol{Q}\boldsymbol{T}\boldsymbol{u}.$$

The following scalar thermodynamic quantities are objective, temperature, pressure, entropy, internal energy, free energy, and enthalpy, etc. However, there are scalars that are clearly not objective such as speed and magnitude of acceleration. Two vectors that are, intuitively, not objective for all observer transformations are velocity and acceleration. For the velocity vector this can be verified by differentiating equation (5.46) once with respect to time to obtain

$$\boldsymbol{v}^* = \boldsymbol{Q}\boldsymbol{v} + \dot{\boldsymbol{c}} + \boldsymbol{\Omega}(\boldsymbol{x}^* - \boldsymbol{c}), \tag{5.54}$$

where $\dot{\boldsymbol{Q}}\boldsymbol{Q}^T = -\boldsymbol{Q}\dot{\boldsymbol{Q}}^T = \boldsymbol{\Omega}$. It is clear that the velocity vector is objective, only for transformations with $\dot{\boldsymbol{c}} = 0$ and $\dot{\boldsymbol{Q}} = 0$. For the acceleration vector, differentiation of equation (5.54) with respect to time and the substitution of $\boldsymbol{v}$ obtained from equation (5.54) gives

$$\boldsymbol{a}^* = \boldsymbol{Q}\boldsymbol{a} + \ddot{\boldsymbol{c}} + \left(\dot{\boldsymbol{\Omega}} - \boldsymbol{\Omega}^2\right)(\boldsymbol{x}^* - \boldsymbol{c}) + 2\boldsymbol{\Omega}(\boldsymbol{v}^* - \dot{\boldsymbol{c}}). \tag{5.55}$$

In equation (5.55) the terms $\dot{\boldsymbol{\Omega}}(\boldsymbol{x}^* - \boldsymbol{c})$, $-\ \boldsymbol{\Omega}^2(\boldsymbol{x}^* - \boldsymbol{c})$, and $2\boldsymbol{\Omega}(\boldsymbol{v}^* - \dot{\boldsymbol{c}})$ are analogous to the terms, which appear in the expressions $a_r = \ddot{r} - r\dot{\theta}^2$, $a_\theta = r\ddot{\theta} + 2\dot{r}\dot{\theta}$, for acceleration components in terms of plane polar coordinates (r, θ). It is clear that the acceleration vector is objective only if

$$\ddot{\boldsymbol{c}} + \left(\dot{\boldsymbol{\Omega}} - \boldsymbol{\Omega}^2\right)(\boldsymbol{x}^* - \boldsymbol{c}) + 2\boldsymbol{\Omega}(\boldsymbol{v}^* - \dot{\boldsymbol{c}}) = \boldsymbol{0},$$

and transformations that satisfy this condition are known as Galilean transformations. A frame of reference, with respect to which Newton's laws are valid, is known as an inertial frame of reference, and all frames obtained from an inertial frame of reference by a Galilean transformation are also inertial frames of reference.

Some further examples of physical quantities and aspects of objectivity of interest in continuum mechanics are now considered. In what follows the objective is taken to mean objective for all observer transformations.

The material time derivative of any objective vector is not objective since differentiation of equation (5.48), with respect to time, gives

$$\dot{\boldsymbol{u}}^* = \boldsymbol{Q}\dot{\boldsymbol{u}} + \dot{\boldsymbol{Q}}\boldsymbol{u}. \tag{5.56}$$

Also the material time derivative,

$$\dot{\boldsymbol{T}} = \frac{\partial \boldsymbol{T}(\boldsymbol{x}, t)}{\partial t} + \boldsymbol{v} \cdot \nabla \boldsymbol{T}(\boldsymbol{x}, t) \tag{5.57}$$

or

$$\dot{\boldsymbol{T}} = \frac{\partial T(x(X)t)}{\partial t},$$

of an objective second-order tensor, $\boldsymbol{T}$ is not objective since it follows from equation (5.52) that

$$\dot{\boldsymbol{T}}^* = \boldsymbol{Q}\dot{\boldsymbol{T}}\boldsymbol{Q}^T + \dot{\boldsymbol{Q}}\boldsymbol{T}\boldsymbol{Q}^T + \boldsymbol{Q}\boldsymbol{T}\dot{\boldsymbol{Q}}^T. \tag{5.58}$$

It is evident that the material rate equation (5.57) is not suitable for use in constitutive equations that involve rates of objective second-order tensors. In order to obtain an objective rate of an objective second-order tensor it is desirable to consider first the two-point deformation gradient tensor, $\boldsymbol{F} = \text{Grad}x$, $(F_{iK} = \partial x_i / X_K)$, and then the velocity gradient tensor, $\boldsymbol{L} = \text{grad}v$, $(L_{ij} = \partial \mathrm{v}_i / \partial x_j)$, and its symmetric and antisymmetric parts. Since no change in the reference configuration of a deforming body is observed by observers on F and F*, Grad $\equiv$ Grad*. It then follows that $\boldsymbol{F}^* = \text{Grad}\boldsymbol{x}^*$ and substitution of $\boldsymbol{x}^* = \boldsymbol{c}(\boldsymbol{t}) + \boldsymbol{Q}(\boldsymbol{t})\boldsymbol{x}(\boldsymbol{X}, t)$ gives

$$\boldsymbol{F}^* = \boldsymbol{QF}. \tag{5.59}$$

According to the condition (5.52), $\boldsymbol{F}$ is not objective; however, it may be regarded as quasi-objective since the reference configuration does not change under an observer transformation. The velocity gradient is given in terms of $\boldsymbol{F}$ by $\boldsymbol{L} = \dot{\boldsymbol{F}}\boldsymbol{F}^{-1}$ and substituting equation (5.59) in $\boldsymbol{L}^* = \dot{\boldsymbol{F}}^* \boldsymbol{F}^{*^{-1}}$ gives

$$\boldsymbol{L}^* = (\dot{\boldsymbol{Q}}\boldsymbol{F} + \boldsymbol{Q}\dot{\boldsymbol{F}})\boldsymbol{F}^{-1}\boldsymbol{Q}^T = \boldsymbol{QLQ}^T + \dot{\boldsymbol{Q}}\boldsymbol{Q}^T, \tag{5.60}$$

which shows that $\boldsymbol{L}$ is not objective. It follows from equation (5.60) that, for the rate of deformation and spin tensors,

$$\boldsymbol{D}^* = \frac{1}{2}\left(\boldsymbol{L}^* + \boldsymbol{L}^{*T}\right) = \boldsymbol{QDQ}^T \tag{5.61}$$

and

$$\boldsymbol{W}^* = \frac{1}{2}\left(\boldsymbol{L}^* - \boldsymbol{L}^{*T}\right) = \boldsymbol{QWQ}^T + \dot{\boldsymbol{Q}}\boldsymbol{Q}^T, \tag{5.62}$$

respectively, which show that the rate of deformation is objective and the spin is not objective.

It is interesting to consider the left and right Cauchy-Green tensors,

$$\boldsymbol{B} = \boldsymbol{FF}^T \quad \text{and} \quad \boldsymbol{C} = \boldsymbol{F}^T\boldsymbol{F}, \tag{5.63}$$

respectively. It can be shown from equations (5.59) and (5.62) that

$$\boldsymbol{B}^* = \boldsymbol{Q}\boldsymbol{B}\boldsymbol{Q}^T \tag{5.64}$$

and

$$\boldsymbol{C}^* = \boldsymbol{C}. \tag{5.65}$$

Equations (5.64) and (5.65) indicate that $\boldsymbol{B}$ is objective but $\mathbf{C}$ does not satisfy condition (5.52). However, like $\boldsymbol{F}$, it may be regarded as quasi-objective since it does not change under an observer transformation. We now consider the material rate of change of a second-order objective tensor. Elimination of $\dot{\boldsymbol{Q}}$ and $\dot{\boldsymbol{Q}}^T$ from equation (5.58),

$$\dot{\boldsymbol{Q}} = \boldsymbol{W}^*\boldsymbol{Q} - \boldsymbol{Q}\boldsymbol{W}$$

and

$$\dot{\boldsymbol{Q}}^T = -\boldsymbol{Q}^T\boldsymbol{W}^* + \boldsymbol{W}\boldsymbol{Q}^T,$$

that are obtained from equation (5.62), gives the relation

$$\left(\dot{\boldsymbol{T}}^* - \boldsymbol{W}\boldsymbol{T}^* + \boldsymbol{T}^*\boldsymbol{W}^*\right) = \boldsymbol{Q}\left(\dot{\boldsymbol{T}} - \boldsymbol{W}\boldsymbol{T} + \boldsymbol{T}\boldsymbol{W}\right)\boldsymbol{Q}^T. \tag{5.66}$$

Consequently,

$$\overset{o}{\boldsymbol{T}} = \left(\dot{\boldsymbol{T}} - \boldsymbol{W}\boldsymbol{T} + \boldsymbol{T}\boldsymbol{W}\right) \tag{5.67}$$

is a possible objective rate. The objective rate equation (5.67) is known as the corotational or Jaumann rate and was proposed by Jaumann in 1911. It is not unique since an infinite number of objective rates can be obtained. Several other objective rates have been proposed for various constitutive relations since 1911, for example,

$$\overset{\Delta}{\boldsymbol{T}} = \left(\dot{\boldsymbol{T}} + \boldsymbol{L}^T\boldsymbol{T} + \boldsymbol{T}\boldsymbol{L}\right) \tag{5.68}$$

is a rate proposed by Oldroyd [7].

The rate equation (5.67) is usually applied to the Cauchy stress tensor $\boldsymbol{\sigma}$ and has a particular physical interpretation that other stress rates do not have, namely, that it is the stress rate relative to a frame of reference rotating with the angular velocity of the principal axes of $\boldsymbol{D}$. To show this let $\boldsymbol{u}_i(t)$ be a right-handed triad of unit base vectors in the principal directions of $\boldsymbol{D}$. The Cauchy stress can be expressed in the form

$$\boldsymbol{\sigma}(t) = \sigma_{ij}(t)\boldsymbol{u}_i(t) \otimes \boldsymbol{u}_j(t),$$

where the material time derivative is

$$\dot{\boldsymbol{\sigma}}(t) = \frac{d}{dt}\left(\boldsymbol{\sigma}_{ij}\mathbf{u}_i \otimes \mathbf{u}_j\right) = \frac{d\boldsymbol{\sigma}_{ij}}{dt}\left(\mathbf{u}_i \otimes \mathbf{u}_j\right) + \boldsymbol{\sigma}_{ij}\dot{\mathbf{u}}_i \otimes \mathbf{u}_j + \boldsymbol{\sigma}_{ij}\mathbf{u}_i \otimes \dot{\mathbf{u}}_j, \tag{5.69}$$

and the σ_{ij} are the stress components with respect to triad $\mathbf{u}_i$, $i \in \{1,2,3\}$. The dependence of $\boldsymbol{\sigma}$ on $\boldsymbol{x}$ has been suppressed. The rate of change of stress relative to the system of base vectors $\mathbf{x}_i(t)$ is $\frac{d\sigma_{ij}}{dt}\left(\mathbf{u}_i \otimes \mathbf{u}_j\right)$. It follows from equation (2.39) that

$$\dot{\mathbf{u}}_i = \boldsymbol{W}\mathbf{u}_i, \quad i \in \{1,2,3\}. \tag{5.70}$$

Substituting equation (5.70) in equation (5.69) and using $\boldsymbol{W} = W_{rs}\mathbf{u}_r \otimes \mathbf{u}_s = -\boldsymbol{W}^T$ gives, after some manipulation,

$$\begin{aligned} \frac{d\sigma_{ij}}{dt}\left(\mathbf{u}_i \otimes \mathbf{u}_j\right) &= \frac{d}{dt}\left(\sigma_{ij}\mathbf{u}_i \otimes \mathbf{u}_j\right) - W_{ip}\boldsymbol{\sigma}_{pj}\mathbf{u}_i \otimes \mathbf{u}_j + \boldsymbol{\sigma}_{ip}W_{pj}\mathbf{u}_i \otimes \mathbf{u}_j \\ \text{or } \overset{o}{\boldsymbol{\sigma}} &= \frac{d\boldsymbol{\sigma}}{dt} - \boldsymbol{W}\boldsymbol{\sigma} + \boldsymbol{\sigma}\boldsymbol{W}. \end{aligned} \tag{5.71}$$

The right-hand sides of equations (5.67), with $\boldsymbol{T}$ replaced by $\boldsymbol{\sigma}$, and (5.71) are equal, so that

$$\overset{o}{\boldsymbol{\sigma}} = \frac{d\boldsymbol{\sigma}_{ij}}{dt}\left(\mathbf{u}_i \otimes \mathbf{u}_j\right).$$

An application of the Oldroyd rate equation (5.68) relates Almansi's strain tensor

$$\eta = \frac{1}{2}\left(I - F^{T^{-1}}F^{-1}\right)$$

to the rate of deformation tensor D since it may be shown that

$$\overset{\Delta}{\eta} = \left(\eta + L^T\eta + \eta L\right) = D.$$

All the terms of materially objective constitutive equations must be objective. For example, in equation (5.9) the quantities σ, D, and p are objective and equation (5.9) is materially objective. However, the constitutive equations for linear elasticity are not materially objective and are valid only for infinitesimal displacement gradients.

EXERCISES

5.1. Decompose the specific strain energy function $W(e)$ for a linear isotropic elastic solid into a dilatational part and a distortional part. Obtain the specific strain energy as a function $W_c(\sigma)$, known as the complementary energy. Decompose $W_c(\sigma)$ into a dilatational part and a distortional part and show that the distortional part is related to the octahedral shearing stress and the second invariant of the deviatoric Cauchy stress.

5.2. The potential energy of a linear elastic body is given by

$$V = \int_{\mathrm{v}} w(e)d\mathrm{v} - \int_{\partial \mathrm{v}_t} t_i u_i d\mathrm{a} - \int_{\mathrm{v}} \rho f_i u_i d\mathrm{v},$$

where $\partial \mathrm{v}_t$ is the part of the surface $\partial \mathrm{v}$ upon which the surface traction is prescribed.

Use the result that the strain energy function, $w(e) = \frac{1}{2}c_{ijkl}e_{ij}e_{kl}$, is positive definite to show that, V is a minimum when $u_i(x)$ is the solution displacement field in competition with all other members of the set of kinematically admissible displacement fields.

Hint, show that $V(u_i^*) - V(u_i) \geqslant 0$, where u_i^* is a kinematically admissible displacement field and u_i is the solution displacement field, with the equality holding if and only if $u_i^* = u_i$

5.3. The complementary energy of a linear elastic body is given by

$$V_c = \int_{\mathrm{v}} w_c(\sigma) dv - \int_{\partial \mathrm{v}_u} t_i u_i da,$$

where $\partial \mathrm{v}_u$ is the part of the surface $\partial \mathrm{v}$ upon which the displacement is prescribed.

Use the result that the complementary energy function, $w_c(\boldsymbol{\sigma}) = \frac{1}{2} a_{ijkl} \sigma_{ij} \sigma_{kl}$, is positive definite, to show that V_c is a minimum when $\sigma_{ij}(\boldsymbol{x})$ is the solution stress field in competition with all other members of the set of statically admissible stress fields.

Hint, show that $V_c\left(\sigma_{ij}^*\right) - V_c(\sigma_{ij}) \geqslant 0$, where $\sigma_{ij}^*(\boldsymbol{x})$ is a statically admissible displacement field and $\sigma_{ij}(x)$ is the solution displacement field, with the equality holding if and only if $\sigma_{ij}^* = \sigma_{ij}$.

5.4. Show that $(\boldsymbol{W} - \Omega)$, where $\boldsymbol{W}$ is the spin tensor, is objective.

5.5. A proposed constitutive equation has the form

$$\dot{\sigma} = 2\mu \boldsymbol{D} + \lambda \boldsymbol{I} \mathrm{tr} \boldsymbol{D},$$

where μ and λ are Lamé's constants of linear elasticity $\dot{\sigma}$ is the material derivative of Cauchy stress. Show that this relation is not materially objective.

5.6. Show that a constitutive of the form

$$\overset{o}{\sigma} = 2\mu \boldsymbol{D} + \lambda \boldsymbol{I} \mathrm{tr} \boldsymbol{D}$$

is materially objective.

5.7. Show that a constitutive equation of the form $\boldsymbol{\sigma} = \boldsymbol{f}(\boldsymbol{D})$ is materially objective.

5.8. Show that the time-dependent basic invariants of the stress tensor σ are stationary when the Jaumann stress rate vanishes.

REFERENCES

1. Müller, I. (1985). Thermodynamics. Pitman.
2. White, F.M. (1991). Viscous Fluid Flow, 2nd Ed. McGraw-Hill.
3. Love, A.E.H. (1927). A Treatise on the Mathematical Theory of Elasticity, 4th Ed. Cambridge University Press.
4. Goodier, J.N., and Timoshenko, S. (1970). Theory of Elasticity, 3rd Ed. McGraw-Hill.
5. Sokolnikov, I.S. (1956). Mathematical Theory of Elasticity, 2nd Ed. McGraw-Hill.
6. Eringen, A.C. (1962). Nonlinear Theory of Continuous Media. McGraw-Hill.
7. Oldroyd, J.G. (1950). On the Formulation of Rheological Equations of State. Proc. Roy. Soc. A 200, pp. 523–541.

6 Finite Deformation of an Elastic Solid

6.1 Introduction

The constitutive equations of the linear theory of elasticity, which have been discussed in chapter 5, can be obtained as a limiting case of those of finite deformation elasticity by neglecting terms $O\left(|\nabla_X \boldsymbol{u}|^2\right)$. However, the constitutive equations of linear elasticity are not objective, although those of finite deformation elasticity are objective. In this chapter we consider the constitutive equations of finite deformation elasticity, henceforth described as nonlinear elasticity. The emphasis is on isotropic solids and it is evident where the specialization to isotropy occurs. Symbolic notation is mainly used, although some equations are given in both symbolic and suffix notation.

6.2 Cauchy Elasticity

An elastic solid is a solid that is rate independent and the components of the Cauchy stress tensor are single-valued functions of the deformation gradient tensor, for either isothermal or adiabatic deformation, so that

$$\boldsymbol{\sigma} = \boldsymbol{\vartheta}(\boldsymbol{F}), \quad \sigma_{ij} = \vartheta_{ij}(F_{mK}). \tag{6.1}$$

In equation (6.1) any dependence on $\boldsymbol{X}$ and t has been suppressed, that is, the solid is assumed to be homogeneous in the undeformed reference configuration. The form of the tensor function $\boldsymbol{\vartheta}$, of the tensor argument $\boldsymbol{F}$, depends on whether the deformation is isothermal or adiabatic and also depends on the reference configuration unless the solid is isotropic. The difference between the forms of $\boldsymbol{\vartheta}$ for isothermal and adiabatic deformation, of materials for which the nonlinear elastic model is realistic, is very small and is negligible for most applications. In this chapter and chapter 7, isothermal deformation is assumed. Equation (6.1) is an example of a constitutive model for a simple material, and the function $\boldsymbol{\vartheta}$ can be obtained from experiments involving homogeneous static deformations.

The function in equation (6.1) must be objective. An objective function is obtained by noting that the stress depends only on the change of shape and is not influenced by rigid body rotations of a deformed configuration. Suppose a deformed configuration of an elastic body is given a rigid rotation, described by a proper orthogonal tensor $\boldsymbol{Q}$, about a fixed material point taken as the origin. The position vector $\boldsymbol{x}$ of a particle before the rotation becomes

$$\boldsymbol{x}^* = \boldsymbol{Q}\boldsymbol{x}$$

after the rotation. Then the deformation gradient and Cauchy stress tensors after rotation are given by,

$$\boldsymbol{F}^* = \boldsymbol{Q}\boldsymbol{F} \tag{6.2}$$

and

$$\boldsymbol{\sigma}^* = \boldsymbol{Q}\boldsymbol{\sigma}\boldsymbol{Q}^T, \tag{6.3}$$

respectively. It follows from equation (6.1) that

$$\boldsymbol{\sigma}^* = \boldsymbol{\vartheta}(\boldsymbol{F}^*). \tag{6.4}$$

Since a rotation of the spatial configuration is equivalent to a rotation of an observer, it further follows from equations (6.1) to (6.4) that $\boldsymbol{\vartheta}$ is observer independent, that is, objective, if

$$\boldsymbol{Q}\boldsymbol{\vartheta}(\boldsymbol{F})\boldsymbol{Q}^T = \boldsymbol{\vartheta}(\boldsymbol{Q}\boldsymbol{F}) \tag{6.5}$$

or

$$\boldsymbol{\vartheta}(\boldsymbol{F}) = \boldsymbol{Q}^T\boldsymbol{\vartheta}(\boldsymbol{Q}\boldsymbol{F})\boldsymbol{Q}.$$

for any $\boldsymbol{Q} \in \text{Orth}^+$ where Orth^+ is the set of all proper orthogonal tensors. Substitution of the polar decomposition of the deformation gradient, $\boldsymbol{F} = \boldsymbol{RU}$, in equation (6.5) gives

$$\boldsymbol{\vartheta}(\boldsymbol{F}) = \boldsymbol{Q}^T \boldsymbol{\vartheta}(\boldsymbol{QRU})\boldsymbol{Q},$$

which is valid for any proper orthogonal tensor, in particular for $\boldsymbol{Q} = \boldsymbol{R}^T$. Consequently,

$$\boldsymbol{\vartheta}(\boldsymbol{F}) = \boldsymbol{R}\boldsymbol{\vartheta}(\boldsymbol{U})\boldsymbol{R}^T, \tag{6.6}$$

which shows that $\boldsymbol{\vartheta}$ can be expressed as a function of six independent components of $\boldsymbol{U}$ instead of nine components of $\boldsymbol{F}$. It follows that equation (6.6) is a necessary condition for equation (6.5) to hold. This condition is also sufficient for by replacing $\boldsymbol{F}$ by $\boldsymbol{QF}$, in equation (6.6), where $\boldsymbol{Q}$ is an arbitrary proper orthogonal tensor, and noting that $(\boldsymbol{QR})\boldsymbol{U}$ is the polar decomposition of $\boldsymbol{QF}$, we obtain

$$\boldsymbol{\vartheta}(\boldsymbol{QF}) = \boldsymbol{QR}\boldsymbol{\vartheta}(\boldsymbol{U})\boldsymbol{R}^T\boldsymbol{Q}^T = \boldsymbol{Q}\boldsymbol{\vartheta}(\boldsymbol{F})\boldsymbol{Q}^T,$$

and this is equivalent to equation (6.5).

The following relation obtained from equations (6.1) and (6.6),

$$\boldsymbol{\sigma} = \boldsymbol{R}\boldsymbol{\vartheta}(\boldsymbol{U})\boldsymbol{R}^T, \tag{6.7}$$

can be put in a different form, which may be convenient since the decomposition of $\boldsymbol{F}$ is not required. Using the polar decomposition of $\boldsymbol{F}$, we obtain

$$\boldsymbol{\sigma} = \boldsymbol{F}\boldsymbol{U}^{-1}\boldsymbol{\vartheta}(\boldsymbol{U})\boldsymbol{U}^{-1}\boldsymbol{F}^T,$$

and then using the right Cauchy Green deformation tensor $\boldsymbol{C} = \boldsymbol{F}^T\boldsymbol{F} = \boldsymbol{U}^2$,

$$\boldsymbol{\sigma} = \boldsymbol{F}\boldsymbol{\phi}(\boldsymbol{C})\boldsymbol{F}^T, \tag{6.8}$$

where

$$\boldsymbol{\phi}(\boldsymbol{C}) = \boldsymbol{C}^{-1/2}\boldsymbol{\vartheta}\left(\boldsymbol{C}^{1/2}\right)\boldsymbol{C}^{-1/2}.$$

It is easily verified that equation (6.8) is objective. Using equation (3.33)

$$\boldsymbol{S} = J\boldsymbol{F}^{-1}\boldsymbol{\sigma}\boldsymbol{F}^{-T},$$

where $\boldsymbol{S}$ is the second Piola-Kirchhoff stress, equation (6.8) can be put in the form

$$\boldsymbol{S} = J\boldsymbol{\phi}(\boldsymbol{C}), \tag{6.9}$$

where $J = \det[\boldsymbol{U}] = \det\left[\sqrt{\boldsymbol{C}}\right] = \det[\boldsymbol{F}]$. The Cauchy stress and the second Piola-Kirchhoff stress can be obtained from equations (6.8) and (6.9), respectively, for a given deformation if the function $\boldsymbol{\phi}$ is known. The function $\boldsymbol{\phi}$ is a symmetric second-order tensor function of a symmetric tensor argument that depends on what is known as the material symmetry and must give rise to a physically reasonable response to the application of stress. For example, application of a hydrostatic pressure should result in a decrease in volume, and an axial tension force applied to a bar should result in an increase in length for either isothermal or isentropic deformation.

According to equations (6.1) and (6.9), the stress in an elastic solid is determined uniquely from the deformation from the undeformed reference configuration and does not depend on the deformation path. A solid that satisfies this condition is described as Cauchy elastic. If the stress cannot be obtained from a scalar potential function known as the strain energy function, the elastic solid is not a conservative system and the work done by the stress could be dependent on the deformation path.

The simplest type of material symmetry is isotropy and the theory for an isotropic elastic solid is given in this chapter. An isotropic material has properties that are directionally independent and this is made more precise in what follows.

If the stress due to an applied deformation is unchanged by a rigid body rotation of the natural reference configuration, the rotation is called a symmetry transformation. The rigid body rotation of the reference configuration from configuration B_1 to B_2 is described by $\boldsymbol{H} \in \text{Orth}^+$ where Orth^+ denotes the set of second-order proper orthogonal tensors, and if the stress is unchanged $\forall \boldsymbol{H}$, where the symbol $\forall$ means for all, the solid is isotropic. Consider a homogeneous deformation described by deformation gradient tensor $\boldsymbol{F}$, referred to the unrotated reference configuration B_1, and by $\boldsymbol{F}'$ referred to B_2. Then with rectangular Cartesian coordinate systems, $(0, \boldsymbol{E}), (0', \boldsymbol{E}')$, related to B_1 and B_2, respectively, and (o,$\mathbf{e}$) related to the spatial configuration,

$$\boldsymbol{F} = \frac{\partial x_i}{\partial X_K} \mathbf{e}_i \otimes \boldsymbol{E}_k, \;\; \boldsymbol{F}' = \frac{\partial x_i}{\partial X'_\alpha} \mathbf{e}_i \otimes \boldsymbol{E}'_\alpha, \text{ and } \boldsymbol{H} = \frac{\partial X'_\beta}{\partial X_K} E'_\beta \otimes E_K,$$

and it follows that $\boldsymbol{F}' = \boldsymbol{F}\boldsymbol{H}^{-1}$. Since $\boldsymbol{F}'$ is the deformation gradient referred to B_2,

$$\boldsymbol{\vartheta}'(\boldsymbol{F}') = \boldsymbol{\vartheta}\left(\boldsymbol{F}\boldsymbol{H}^{-1}\right),$$

where $\boldsymbol{\vartheta}'$ is the response function relative to B_2. An isotropic solid has a response function that does not depend on rotation of the reference configuration so that for isotropy $\boldsymbol{\vartheta}' = \boldsymbol{\vartheta}$, and

$$\boldsymbol{\vartheta}(\boldsymbol{F}) = \boldsymbol{\vartheta}(\boldsymbol{F}\boldsymbol{P}), \quad \text{for} \quad \forall \boldsymbol{P} \in \text{Orth}^+, \tag{6.10}$$

where $\boldsymbol{P} = \boldsymbol{H}^{-1} = \boldsymbol{H}^T$. The set orth^+ is the symmetry group for an isotropic elastic solid. A group is a set G that satisfies the following conditions: (1) an operation (multiplication) is defined on G, that is, the product of two members of the set is also a member; (2) the set G has a unit element; and (3) every member of G has an inverse in G. It may be verified that Orth^+ is a group. If equation (6.10) is not true for all $\boldsymbol{P} \in \text{Orth}^+$ but is true for a subgroup of Orth^+, the solid is anisotropic and the subgroup is known as the symmetry group of the solid. Finite deformation of anisotropic elastic solids is not considered in this text and details can be found in [1].

The function $\boldsymbol{\phi}(\boldsymbol{C})$ in equation (6.9) is an isotropic function of $\boldsymbol{C}$, that is, by definition, $\boldsymbol{Q}\boldsymbol{\phi}(\boldsymbol{C})\boldsymbol{Q}^T = \boldsymbol{\phi}(\boldsymbol{Q}\boldsymbol{C}\boldsymbol{Q}^T), \forall \boldsymbol{Q} \in \text{Orth}^+$ if the solid is isotropic.

It may be shown [2] that $\boldsymbol{\phi}$ is an isotropic function if and only if it can be expressed in the form

$$\boldsymbol{\phi} = \phi_0 \boldsymbol{I} + \phi_1 \boldsymbol{C} + \phi_2 \boldsymbol{C}^2, \tag{6.11}$$

where $\phi_o, \phi_1,$ and ϕ_2 are scalar functions of the basic invariants of $\boldsymbol{C}$,

$$I_1 = \mathrm{tr}\boldsymbol{C}, \quad I_2 = \frac{1}{2}\left\{(\mathrm{tr}\boldsymbol{C})^2 - \mathrm{tr}\boldsymbol{C}^2\right\}, \quad I_3 = \det \boldsymbol{C}.$$

The invariants of $\boldsymbol{C}$, as the term indicates, are unchanged if $\boldsymbol{C}$ is replaced by $\boldsymbol{QCQ}^T$ for $\forall \boldsymbol{Q} \in \mathrm{ortho}^+$.

If the deformation is given and $\phi_0, \phi_1,$ and ϕ_2 as functions of I_1, I_2 and, I_3 are known, the second Piola-Kirchhoff stress can be obtained from equations (6.9) and (6.11) for an isotropic solid.

6.3 Hyperelasticity

If a strain energy function, often referred to as a stored energy function w, exists, the solid is known as Green elastic or hyperelastic.

The class of hyperelastic solids is a subclass of the class of Cauchy elastic solids, and in this book we assume that elastic materials are hyperelastic. Strain energy functions correspond to the Helmholtz free energy for isothermal deformation and the internal energy for isentropic deformation.

It has been shown in chapter 4 that the rate of work done by the body forces and surface tractions acting on a body is equal to the sum of the integrated stress power of the body and rate of change of the kinetic energy of the body. It may be deduced from this that, if the system is conservative, the integrated stress power is equal to the rate of change of stored energy. This motivates the definition of a hyperelastic solid that it is a solid for which the specific stress power is equal to the rate of change of the specific strain energy [3]. The specific stress power and specific strain energy, henceforth called the stress power and strain energy function, respectively, are per unit volume of the undeformed reference configuration in this chapter, rather than per unit mass as is customary for specific thermodynamic properties. The specific strain energy w for finite deformation of a hyperelastic solid can conveniently be expressed as a function of one of, $E, C, U,$ or F. A dependence on $\boldsymbol{X}$ is suppressed so that it is assumed that the solid is homogeneous in the reference configuration. The function w must be unchanged by a rotation of the deformed configuration. This condition is satisfied if w is expressed as a function of $\boldsymbol{E}, \boldsymbol{C},$ or $\boldsymbol{U}$. If it is expressed as a function of $\boldsymbol{F}$, then the condition $w(\boldsymbol{F}) = w(\boldsymbol{QF})$ must be satisfied $\forall \boldsymbol{Q} \in \text{Orth}^+$ in particular for $\boldsymbol{Q} = \boldsymbol{R}^T$. It then follows that $w(\boldsymbol{F}) = w(\boldsymbol{U})$. Since the Green strain tensor $\boldsymbol{E} = \frac{1}{2}(\boldsymbol{U}^2 - \boldsymbol{I})$, and the Green deformation tensor $\boldsymbol{C} = \boldsymbol{U}^2$, the strain energy function must also be expressible in the forms $w(\boldsymbol{E})$ and $w(\boldsymbol{C})$ where we note that at this stage the same symbol, w, has been used for different functions of different arguments that represent the same quantity.

In order to obtain the stress strain relations for a hyperelastic solid we first consider the stress power in terms of the second Piola-Kirchhoff stress tensor $\boldsymbol{S}$ and the Green strain rate tensor $\dot{\boldsymbol{E}}$ so that the specific stress power P, is given by

$$P = \text{tr}(\boldsymbol{S}\dot{\boldsymbol{E}}) = S_{KL}\dot{E}_{KL}, \tag{6.12}$$

and $P = \dot{w}(\boldsymbol{E})$. It is convenient to use suffix notation for the next development. Then

$$\dot{w}(\boldsymbol{E}) = \frac{\partial w(\boldsymbol{E})}{\partial E_{KL}} \dot{E}_{KL}. \tag{6.13}$$

It should be noted that in equation (6.13) all components of E_{KL} are taken as independent, for the differentiation, that is, the symmetry of E_{KL} is ignored. Then it follows from equations (6.12) and (6.13) that

$$S_{LK} = \frac{\partial w(\boldsymbol{E})}{\partial E_{KL}}. \tag{6.14}$$

In order to relate to equation (6.9), we now consider the stress power in terms of the second Piola-Kirchhoff stress tensor $\boldsymbol{S}$ and the time rate of the right Green deformation tensor $\boldsymbol{C}$. It then follows from equation (6.12) and $\boldsymbol{E} = (1/2)(\boldsymbol{C} - \boldsymbol{I})$ that

$$P = \frac{1}{2}\text{tr}\boldsymbol{S}\dot{\boldsymbol{C}} = \frac{1}{2}S_{KL}\dot{C}_{KL} \tag{6.15}$$

and

$$S_{LK} = 2\frac{\partial w(\boldsymbol{C})}{\partial C_{KL}}. \tag{6.16}$$

Equations (6.14) and (6.16) are unchanged by rotation of the deformed configuration.

It follows from equations (6.9), (6.11), and (6.16) that, for an isotropic elastic solid,

$$2\frac{\partial w(\boldsymbol{C})}{\partial C_{KL}} = J\left(\phi_0 + \phi_1\boldsymbol{C} + \phi_2\boldsymbol{C}^2\right),$$

and this determines $w(\boldsymbol{C})$. For isotropy, the strain energy function $w(\boldsymbol{C})$ is a scalar isotropic function of $\boldsymbol{C}$, that is,

$$w\left(\boldsymbol{Q}\boldsymbol{C}\boldsymbol{Q}^T\right) = w(\boldsymbol{C}).$$

This condition is satisfied if w is expressed as a function,

$$w = \hat{w}(I_1, I_2, I_3), \tag{6.17}$$

of the invariants of $\boldsymbol{C}$. Since the strain energy must be zero for no deformation w must satisfy the equivalent conditions,

$$w(\boldsymbol{I}) = 0 \quad \text{or} \quad \hat{w}(3,3,1) = 0.$$

The form (6.17) has been used extensively in the literature of hyperelasticity, and constitutive equations based on this form are now developed as follows:

$$\frac{\partial w(\boldsymbol{C})}{\partial C_{KL}} = \sum_{i=1}^{3}\frac{\partial \hat{w}}{\partial I_i}\frac{\partial I_i}{\partial C_{KL}}, \tag{6.18}$$

where

$$\frac{\partial I_1}{\partial C_{KL}} = \delta_{KL}, \ \frac{\partial I_2}{\partial C_{KL}} = I_1\delta_{KL} - C_{KL}, \ \frac{\partial I_3}{\partial C_{KL}} = I_2\delta_{KL} - I_1 C_{KL} + C_{KM}C_{ML.} \tag{6.19}$$

Equations (6.16), (6.18), and (6.19) then give

$$S_{KL} = 2\left[\left(\frac{\partial \hat{w}}{\partial I_1} + I_1\frac{\partial \hat{w}}{\partial I_2} + I_2\frac{\partial \hat{w}}{\partial I_3}\right)\delta_{KL} - \left(\frac{\partial \hat{w}}{\partial I_2} + I_1\frac{\partial \hat{w}}{\partial I_3}\right)C_{KL} + \frac{\partial \hat{w}}{\partial I_3}C_{KM}C_{MK}\right], \tag{6.20}$$

$$\boldsymbol{S} = 2\left[\left(\frac{\partial \hat{w}}{\partial I_1} + I_1\frac{\partial \hat{w}}{\partial I_2} + I_2\frac{\partial \hat{w}}{\partial I_3}\right)\boldsymbol{I} - \left(\frac{\partial \hat{w}}{\partial I_2} + I_1\frac{\partial \hat{w}}{\partial I_3}\right)\boldsymbol{C} + \frac{\partial \hat{w}}{\partial I_3}\boldsymbol{C}^2\right].$$

The relation between the Cauchy stress and the second Piola-Kirchhoff stress

$$\boldsymbol{\sigma} = J^{-1}\boldsymbol{FSF}^T$$

has already been obtained in chapter 5. Using this relation and $\boldsymbol{FCF}^T = \boldsymbol{B}^2$ and $\boldsymbol{FC}^2\boldsymbol{F}^T = \boldsymbol{B}^3$, gives

$$\boldsymbol{\sigma} = 2J^{-1}\left[\left(\frac{\partial \hat{w}}{\partial I_1} + I_1\frac{\partial \hat{w}}{\partial I_2} + I_2\frac{\partial \hat{w}}{\partial I_3}\right)\boldsymbol{B} - \left(\frac{\partial \hat{w}}{\partial I_2} + I_1\frac{\partial \hat{w}}{\partial I_3}\right)\boldsymbol{B}^2 + \frac{\partial \hat{w}}{\partial I_3}\boldsymbol{B}^3\right],$$

and by using the Cayley-Hamilton theorem to eliminate $\boldsymbol{B}^3$ we obtain

$$\boldsymbol{\sigma} = 2J^{-1}\left\{I_3\frac{\partial \hat{w}}{\partial I_3}\boldsymbol{I} + \left(\frac{\partial \hat{w}}{\partial I_1} + I_1\frac{\partial \hat{w}}{\partial I_2}\right)\boldsymbol{B} - \frac{\partial \hat{w}}{\partial I_2}\boldsymbol{B}^2\right\}. \tag{6.21}$$

We note that the invariants of $\boldsymbol{B}$ are the same as those of $\boldsymbol{C}$. An alternative form of equation (6.21),

$$\boldsymbol{\sigma} = 2J^{-1}\left\{\left(I_2\frac{\partial \hat{w}}{\partial I_2} + I_3\frac{\partial \hat{w}}{\partial I_3}\right)\mathrm{I} + \frac{\partial \hat{w}}{\partial I_1}\boldsymbol{B} - I_3\frac{\partial \hat{w}}{\partial I_2}\boldsymbol{B}^{-1}\right\}, \tag{6.21a}$$

is obtained by eliminating $\boldsymbol{B}^2$ from equation (6.21) and $\boldsymbol{B}^2 = I_1\boldsymbol{B} - I_2\boldsymbol{I} + I_3\boldsymbol{B}^{-1}$ that follows from the Cayley-Hamilton theorem. It is evident from equations (6.20) and (6.21) that, for an isotropic solid, the tensors $\boldsymbol{S}$ and $\boldsymbol{C}$ and $\boldsymbol{\sigma}$ and $\boldsymbol{B}$ are coaxial.

Since $\boldsymbol{\sigma} = \boldsymbol{0}$ when $\boldsymbol{B} = \boldsymbol{C} = \boldsymbol{I}$, or $(I_1, I_2, I_3) = (3, 3, 1)$, it follows from equation (6.21) that the strain energy function $\hat{w}(I_1, I_2, I_3)$ must satisfy the condition

$$\left(\frac{\partial \hat{w}}{\partial I_1} + 2\frac{\partial \hat{w}}{\partial I_2} + \frac{\partial \hat{w}}{\partial I_3}\right)_{\boldsymbol{C}=\boldsymbol{B}=\boldsymbol{I}} = 0,$$

in addition to $\hat{w}(3,\ 3,\ 1) = 0$. Examples of physically realistic strain energy functions are given in chapter 7.

The equality of the specific stress power and the specific strain energy rate can also be applied to the nominal stress and the time rate of the deformation gradient tensor to obtain

$$\dot{w}(\boldsymbol{F}) = \mathrm{tr}\big(\boldsymbol{\Sigma}\dot{\boldsymbol{F}}\big). \tag{6.22}$$

It then follows from

$$\dot{w}(\boldsymbol{F}) = \frac{\partial w(\boldsymbol{F})}{\partial F_{IK}}\dot{F}_{IK}$$

and equation (6.22)

$$\Sigma_{Ki} = \frac{\partial w(\boldsymbol{F})}{\partial F_{iK}}. \tag{6.23}$$

The relations equations (6.20) and (6.21) can be obtained from equation (6.23), and this is left as an exercise for the reader.

The pairs of tensors $\{\boldsymbol{S},\ \boldsymbol{E}\,(\text{or}\,\boldsymbol{C}/2)\}$, $\{\boldsymbol{\Sigma},\ \boldsymbol{F}\}$ are said to be conjugate since the stress power relations (6.12) or (6.15) and (6.22) are the trace of $\boldsymbol{S}$ and the time rate of $\boldsymbol{E}$ (or $\boldsymbol{C}/2$) and the trace of $\boldsymbol{\Sigma}$ and the time rate of $\boldsymbol{F}$. The stress tensor $\boldsymbol{\sigma}$ and the rate of deformation tensor $\boldsymbol{D}$, which appear in the stress power per unit reference volume, relation, $P = J\mathrm{tr}(\boldsymbol{\sigma}\boldsymbol{D})$, do not give rise to conjugate variables since $\boldsymbol{D}$ is not the time derivative of a strain except for infinitesimal strains where $\boldsymbol{D} \simeq \dot{\boldsymbol{e}}$, and e is the infinitesimal strain tensor.

6.4 Incompressible Hyperelastic Solid

An incompressible elastic solid may be defined as an elastic solid that can undergo only isochoric deformation, that is, the deformation is independent of the hydrostatic part of the stress tensor. This implies that the bulk modulus is infinite so that the solid is mechanically incompressible. Müller [4] has shown that a mechanically incompressible fluid must have zero coefficient of volume thermal expansion, and this proof can be extended to consider a mechanically incompressible solid. It then follows that the volume coefficient of thermal expansion is zero. Most solid rubbers, as opposed to foam rubbers, have bulk moduli that are orders of magnitude greater than their shear moduli so that for most applications volume deformation can be neglected compared with distortional deformation and the incompressible model is applicable. The incompressible model is not applicable when the propagation of longitudinal waves is considered. It should be noted, however, that solid rubber-like materials have volume coefficients of thermal expansion that are large compared with those for metals and several other solids.

The stresses in an incompressible solid can be obtained from the deformation only to within an isotropic Cauchy stress that must be obtained from the stress boundary conditions. Since the invariant $I_3 = 1$ for incompressible materials the strain energy functions for incompressible isotropic hyperelastic solids are of the form

$$w = \hat{w}(I_1, I_2). \tag{6.24}$$

Stress-deformation relations, for an incompressible hyperelastic solid, are obtained by introducing the Lagrangian multiplier p and using the equation of constraint,

$$I_3 - 1 = \det \boldsymbol{C} - 1 = 0. \tag{6.25}$$

Equation (6.16) is then replaced by

$$S_{LK} = 2\frac{\partial w(\boldsymbol{C})}{\partial C_{KL}} - p\frac{\partial \det \boldsymbol{C}}{\partial C_{KL}}. \tag{6.26}$$

It follows from equation (1.74) that

$$C_{IJ}C^{c}_{KJ} = \delta_{IK} \det \boldsymbol{C},$$

$$\boldsymbol{C}\boldsymbol{C}^{cT} = \boldsymbol{I}\det\boldsymbol{C},$$

where C^{c}_{KJ} is the cofactor of C_{KJ}, and it may be deduced that

$$\frac{\partial \det \boldsymbol{C}}{\partial C_{KL}} = C^{c}_{KL} = C^{-1}_{KL} \det \boldsymbol{C}. \tag{6.27}$$

Then from equations (6.25), (6.27), and the method of Lagrangian multipliers we obtain

$$\frac{\partial \det \boldsymbol{C}}{\partial C_{KL}} = C^{-1}_{KL}. \tag{6.28}$$

Substitution of equation (6.28) in equation (6.26) gives

$$S_{LK} = 2\frac{\partial \hat{w}(\boldsymbol{C})}{\partial C_{KL}} - pC^{-1}_{KL}, \tag{6.29}$$

for an incompressible hyperelastic solid. Since $I_3 = 1$ and $w = \hat{w}(I_1, I_2)$ for incompressibility, equation (6.18) becomes

$$\frac{\partial w(\mathbf{C})}{\partial C_{KL}} = \sum_{i=1}^{2} \frac{\partial \hat{w}}{\partial I_i}\frac{\partial I_i}{\partial C_{KL}},$$

where

$$\frac{\partial I_1}{\partial C_{KL}} = \delta_{KL}, \quad \frac{\partial I_2}{\partial C_{KL}} = I_1\delta_{KL} - C_{KL},$$

and substituting in equation (6.29) gives

$$S_{KL} = 2\left[\left(\frac{\partial \hat{w}}{\partial I_1} + I_1 \frac{\partial \hat{w}}{\partial I_2}\right)\delta_{KL} - \frac{\partial \hat{w}}{\partial I_2} C_{KL}\right] - pC_{KL}^{-1}, \tag{6.30}$$

$$\mathbf{S} = 2\left[\left(\frac{\partial \hat{w}}{\partial I_1} + I_1 \frac{\partial \hat{w}}{\partial I_2}\right)\mathbf{I} - \frac{\partial \hat{w}}{\partial I_2}\mathbf{C}\right] - p\mathbf{C}^{-1}.$$

The relation between $\boldsymbol{\sigma}$ and $\boldsymbol{S}$ for an incompressible solid is $\boldsymbol{\sigma} = \boldsymbol{F}\boldsymbol{S}\boldsymbol{F}^T$ and applying this to equation (6.30) gives

$$\sigma_{ij} = 2\left[\left(\frac{\partial \hat{w}}{\partial I_1} + I_1 \frac{\partial \tilde{w}}{\partial I_2}\right)B_{ij} - \frac{\partial \hat{w}}{\partial I_2} B_{ik}B_{kj}\right] - p\delta_{ij}, \tag{6.31}$$

$$\boldsymbol{\sigma} = 2\left[\left(\frac{\partial \hat{w}}{\partial I_1} + I_1 \frac{\partial \hat{w}}{\partial I_2}\right)\boldsymbol{B} - \frac{\partial \hat{w}}{\partial I_2}\boldsymbol{B}^2\right] - p\boldsymbol{I}.$$

Equation (6.31) can be put in a different form

$$\boldsymbol{\sigma} = 2\left[\frac{\partial \hat{w}}{\partial I_1}\boldsymbol{B} - \frac{\partial \hat{w}}{\partial I_2}\boldsymbol{B}^{-1}\right] - p\boldsymbol{I}, \tag{6.32}$$

by using the Cayley-Hamilton theorem to eliminate $\boldsymbol{B}^2$. It should be noted that the Lagrangian multiplier p in equation (6.32) is different from that in the previous equations and that it is not determined by the deformation but by stress boundary conditions.

The relation (6.23) for the nominal stress is replaced by

$$\Sigma_{Ki} = \frac{\partial w(\boldsymbol{F})}{\partial F_{iK}} - pF_{iK}^{-1}, \tag{6.33}$$

for the incompressible model.

6.5 Alternative Formulation

For some problems involving isotropic hyperelastic solids there is an advantage in expressing the strain energy as a symmetric function

$$w = \breve{w}(\lambda_1, \lambda_2, \lambda_3) = \breve{w}(\lambda_2, \lambda_1, \lambda_3) = \breve{w}(\lambda_2, \lambda_3, \lambda_1), \tag{6.34}$$

of the principal stretches λ_i, $i \in \{1,2,3\}$, which are the eigenvalues of $\boldsymbol{U} = \sqrt{\boldsymbol{C}}$ or $\boldsymbol{V} = \sqrt{\boldsymbol{B}}$. It is evident that

$$\breve{w}(1,1,1) = 0.$$

If the eigenvalues of $\boldsymbol{C}$ are distinct it can be expressed as the spectral decomposition

$$\boldsymbol{C} = \sum_{i=1}^{3} \lambda_i^2 \boldsymbol{l}^{(i)} \otimes \boldsymbol{l}^{(i)},$$

where the unit vectors $\boldsymbol{l}^{(i)}, i \in \{1,2,3\}$ are the eigenvectors of $\boldsymbol{C}$ and $\boldsymbol{U}$, corresponding to the λ_i. If only two eigenvalues of $\sqrt{\boldsymbol{C}}$, λ_1 and λ_2, are distinct and $\lambda_2 = \lambda_3$, the spectral decomposition of $\boldsymbol{C}$ is

$$\boldsymbol{C} = \lambda_1^2 \boldsymbol{l}^{(1)} \otimes \boldsymbol{l}^{(1)} + \lambda_2^2 \left(\boldsymbol{I} - \boldsymbol{l}^{(1)} \otimes \boldsymbol{l}^{(1)} \right),$$

and if $\lambda = \lambda_1 = \lambda_2 = \lambda_3$, the spectral decomposition of $\boldsymbol{C}$ is

$$\boldsymbol{C} = \lambda^2 \boldsymbol{I}.$$

The spectral decomposition of a second-order symmetric tensor is also considered in chapter 1.

Principal invariants of $\boldsymbol{C}$ or $\boldsymbol{B}$, in terms of the λ_i, are

$$I_1 = \lambda_1^2 + \lambda_2^2 + \lambda_3^2, \quad I_2 = \lambda_1^2\lambda_2^2 + \lambda_2^2\lambda_3^2 + \lambda_3^2\lambda_1^2, \quad I_3 = \lambda_1^2\lambda_2^2\lambda_3^2,$$

and substitution in equation (6.17) gives equation (6.34).

It follows from $\boldsymbol{C} = \boldsymbol{U}^2$, $\boldsymbol{S}^B = \boldsymbol{SU} = \boldsymbol{US}$ and equation (6.15) that the stress power P can be expressed as

$$P = \mathrm{tr}\left(\boldsymbol{S}^B \dot{\boldsymbol{U}} \right),$$

so that S^B and $\boldsymbol{U}$ are conjugate. Then, since $\dot{w}(\boldsymbol{U}) = \mathrm{tr}((\partial w / \partial \boldsymbol{U}) \dot{\boldsymbol{U}})$,

$$\boldsymbol{S}^B = \frac{\partial \breve{w}(\boldsymbol{U})}{\partial \boldsymbol{U}}. \tag{6.35}$$

The eigenvalues of $\boldsymbol{U}$ are $(\lambda_1, \lambda_2, \lambda_3)$ and it follows from equation (6.35) that the principal components of the Biot stress are then given by

$$S_i^B = \frac{\partial w(\lambda_1, \lambda_2, \lambda_3)}{\partial \lambda_i}, \quad i \in \{1,2,3\}, \tag{6.36}$$

and it follows that

$$\frac{\partial w(1,1,1)}{\partial \lambda_i} = 0, \quad i \in \{1,2,3\}.$$

It may be shown from, $\boldsymbol{\sigma} = J^{-1}\boldsymbol{FSF}^T, \boldsymbol{F} = \boldsymbol{RU}$ and the Biot stress given by

$$\boldsymbol{S}^B = \boldsymbol{US} = \boldsymbol{SU},$$

that

$$J\boldsymbol{R}^T\boldsymbol{\sigma}\boldsymbol{R} = \boldsymbol{S}^B\boldsymbol{U}. \tag{6.37}$$

The tensor $J\boldsymbol{R}^T\boldsymbol{\sigma}\boldsymbol{R}$ and the weighted Cauchy stress $J\boldsymbol{\sigma}$ have the same eigenvalues values. Consequently, it follows from equation (6.36) and the principal values of each side of equation (6.37) that the principal components of Cauchy stress are given by

$$\sigma_i = J^{-1}\lambda_i \frac{\partial \breve{w}}{\partial \lambda_i}, \quad i \in \{1,2,3\}. \text{No summation.} \tag{6.38}$$

This result can be obtained otherwise. Let $L_1, L_2,$ and L_3 be the lengths of the contiguous sides of a rectangular block in the reference configuration. Apply an irrotational, homogeneous deformation such that the lengths of the corresponding sides and stretches become $l_1, l_2,$ and l_3 and $\lambda_1, \lambda_2,$ and λ_3, respectively. Application of the principle of virtual work gives

$$L_1L_2L_3\delta\breve{w}(\lambda_1, \lambda_2, \lambda_3) = \sigma_1 l_2 l_3 \delta l_1 + \sigma_2 l_1 l_3 \delta l_2 + \sigma_3 l_1 l_2 \delta l_3,$$

and, since $\delta\breve{w} = \sum_i^3 \frac{\partial \breve{w}}{\partial \lambda_i}\delta\lambda_i$, $J = \frac{l_1 l_2 l_3}{L_1 L_2 L_3}$, and the $\delta\lambda_i$ are arbitrary the result (6.38) follows.

It should be noted that, for an isotropic elastic solid, the right stretch tensor, $\boldsymbol{U}, \boldsymbol{S}$, and $\boldsymbol{S}^B$ are coaxial and the left stretch tensor, $\boldsymbol{V}$, and $\boldsymbol{\sigma}$ are coaxial.

The use of the strain energy function in the form (6.34) along with equation (6.36) or (6.38) could be the preferred method of solution for problems with the principal directions of $\boldsymbol{\sigma}$ and/or $\boldsymbol{S}^B$ known beforehand. Problems that involve cylindrical or spherical symmetry are examples. For problems such as simple shear and the torsion of a cylindrical bar the use of form (6.34) is less desirable since the principal directions of stress have to be found initially.

For incompressibility, $\lambda_1\lambda_2\lambda_3 = 1$. An assumption, known as the Valanis-Landel hypothesis [5], for an incompressible isotropic elastic solid is

$$\breve{w}(\lambda_1, \lambda_2, \lambda_3) = \tilde{w}(\lambda_1) + \tilde{w}(\lambda_2) + \tilde{w}(\lambda_3),$$

where $\tilde{w}(1) = 0$ and $\tilde{w}'(1) + w''(1) = 2\mu$. Most strain energy functions that have been proposed for incompressible isotropic elastic solids satisfy this hypothesis.

For incompressibility equations (6.36) and (6.38) are replaced by

$$S_i^B = \frac{\partial \breve{w}}{\partial \lambda_i} - \frac{p}{\lambda_i}, \quad i \in \{1,2,3\}. \text{No summation},$$

and

$$\sigma_i = \lambda_i \frac{\partial \breve{w}}{\partial \lambda_i} - p, \quad i \in \{1,2,3\}. \text{No summation}, \tag{6.39}$$

respectively.

Since $\lambda_1\lambda_2\lambda_3 = 1$, it is possible to eliminate λ_3 from $\breve{w}$ to obtain

$$\bar{w}(\lambda_1, \lambda_2) = \breve{w}\left(\lambda_1, \lambda_2, \lambda_1^{-1}\lambda_2^{-1}\right). \tag{6.40}$$

It follows from equation (6.40) and $\lambda_3 = (\lambda_1\lambda_2)^{-1}$ that

$$\frac{\partial \bar{w}}{\partial \lambda_1} = \frac{\partial \breve{w}}{\partial \lambda_1} + \frac{\partial \breve{w}}{\partial \lambda_3}\frac{\partial \lambda_3}{\partial \lambda_1} = \frac{\partial \breve{w}}{\partial \lambda_1} - \lambda_1^{-2}\lambda_2^{-1}\frac{\partial \breve{w}}{\partial \lambda_3}. \tag{6.41}$$

Then it may be deduced from equations (6.39) and (6.41) that

$$\sigma_1 - \sigma_3 = \lambda_1 \frac{\partial \bar{w}}{\partial \lambda_1}. \tag{6.42}$$

Similarly

$$\sigma_2 - \sigma_3 = \lambda_2 \frac{\partial \bar{w}}{\partial \lambda_2}. \tag{6.43}$$

Equations (6.42) and (6.43) are useful when plane stress problems with $\sigma_3 = 0$ are considered.

Further aspects of strain energy functions are given in chapters 7 and 8.

EXERCISES

6.1. Show that the basic invariants of the left Cauchy-Green tensor $\boldsymbol{B}$ are of the same form as those of the right Cauchy-Green tensor $\boldsymbol{C}$, that is,

$$I_1 = \mathrm{tr}\boldsymbol{B}, \quad I_2 = \frac{1}{2}\left\{(\mathrm{tr}\boldsymbol{B})^2 - \mathrm{tr}\boldsymbol{B}^2\right\}, \quad I_3 = \det \boldsymbol{B}.$$

6.2. Show that for consistency with classical linear elastic theory the strain energy function, $w(\lambda_1, \lambda_2, \lambda_3)$, must satisfy the condition

$$\frac{\partial^2 w}{\partial \lambda_i \partial \lambda_j}(1,1,1) = \lambda + 2\mu\delta_{ij},$$

where λ and μ are Lamé's constants of linear classical elasticity.

6.3. The Mooney-Rivlin strain energy function for an incompressible hyperelastic solid is given by

$$w = \frac{\mu}{2}\{\alpha(I_1 - 3) + (1-\alpha)(I_2 - 3)\},$$

where $0 < \alpha \leqslant 1$. Show that it can be put in the form

$$\breve{w}(\lambda_1, \lambda_2, \lambda_3) = \tilde{w}(\lambda_1) + \tilde{w}(\lambda_2) + \tilde{w}(\lambda_3),$$

where $\tilde{w}(1) = 0$ and $\tilde{w}'(1) + w''(1) = 2\mu$.

6.4 Show that, for an incompressible neoHookean elastic solid, the constitutive equation $\sigma = -p\boldsymbol{I} + \mu\boldsymbol{B}$ is materially objective.

REFERENCES

1. Ogden, R.W. (1984). Non-Linear Elastic Deformations. Ellis Harwood.
2. Rivlin, R.S., and Ericksen, J.L. (1955). Stress-Deformation Relations for Isotropic Materials. J. Rat. Mech. Anal. 4, pp. 329–422.
3. Haupt, P. (2000). Continuum Mechanics and Theory of Materials. Springer.
4. Müller, I. (1985). Thermodynamics. Pitman.
5. Valanis, K.C., and Landel, R.S. (1967). The Strain-Energy Function of a Hyperelastic Material in Terms of the Extension Ratios. J. Appl. Phys. 38, pp. 2997–3002.

7 Some Problems of Finite Elastic Deformation

7.1 Introduction

In this chapter problems of finite deformation elastostatics for isotropic hyperelastic solids are considered. Exact solutions for some problems of finite deformation of incompressible elastic solids have been obtained by inverse methods. In these methods a deformation field is assumed, and it is verified that the equilibrium equations and stress and displacement boundary conditions are satisfied. There are some elastostatic problems that have deformation fields that are possible in every homogeneous isotropic incompressible elastic body in the absence of body forces. These deformation fields are said to be controllable. In an important paper by Ericksen [1] an attempt is made to obtain all static deformation fields that are possible in all homogeneous isotropic incompressible bodies acted on by surface tractions only. Ericksen further showed that the only deformation possible in all homogeneous compressible elastic bodies, acted on by surface tractions only, is a homogeneous deformation [2]. In this chapter solutions are given for several problems of finite deformation of incompressible isotropic hyperelastic solids and two simple problems for an isotropic compressible solid. These solutions with one exception involve controllable deformation fields and are based on the physical components of stress, deformation gradient, and the left Cauchy-Green strain tensor. This is in contrast to the approach of Green and Zerna [3], who used convected coordinates and tensor components.

Rivlin, in the late 1940's, was the first to obtain most of the existing solutions, for deformation of an isotropic incompressible hyperelastic solid. Several important papers by Rivlin, containing these solutions, are reproduced in a volume edited by Truesdell [4].

7.2 Strain Energy Functions and Stress-Strain Relations

Solutions are given in this chapter for an incompressible solid based on the Mooney-Rivlin strain energy function,

$$w(I_1, I_2) = \frac{\mu}{2}\{\gamma(I_1 - 3) + (1 - \gamma)(I_2 - 3)\}, \tag{7.1}$$

where μ is the shear modulus for infinitesimal deformation from the natural reference state and $0 < \gamma \leqslant 1$ is a constant, or its special case with $\gamma = 1$, the neo-Hookean strain energy function

$$w(I_1) = \frac{\mu}{2}(I_1 - 3). \tag{7.2}$$

It has been found experimentally that, for simple tension, equations (7.1), with $\gamma = 0.6$, and (7.2) give results in good results for rubber-like materials for axial stretches up to about 3.0 and 1.8, respectively. For larger stretches a strain energy function of the form $w = \tilde{w}(\lambda_1, \lambda_2, \lambda_3)$, which is not readily put in the form (6.24), gives results for simple tension that are in good agreement with experiment. This strain energy function is considered in the final section of this chapter.
Other strain energy functions for isotropic incompressible solids are discussed in [5].

The Hadamard strain energy function

$$w(I_1, I_2, I_3) = \frac{\mu}{2}\{\gamma(I_1 - 3) + (1 - \gamma)(I_2 - 3) + H(I_3)\}, \tag{7.3}$$

where $0 < \gamma \leqslant 1$, $H(1) = 0$ and $3 - \gamma + H'(1) = 0$, is a compressible generalization of the Mooney-Rivlin. A compressible generalization of the neo-Hookean strain energy function (7.2),

$$w(I_1, I_3) = \frac{\mu}{2}\left\{(I_1 - 3) + \frac{1 - 2v}{v}\left(I_3^{\frac{-v}{1-2v}} - 1\right)\right\}, \tag{7.4}$$

has been proposed by Blatz and Ko [6]. In equation (7.4), v is Poisson's ratio for infinitesimal deformation from the natural reference state, and as

$\nu \to 0.5$ and $I_3 \to 1$, (7.4) $\to$ (7.2). Several other compressible generalizations of equations (7.1) and (7.2) have been proposed [7].

Hyperelastic stress-deformation relations used in this chapter are

$$\boldsymbol{\sigma} = -p\boldsymbol{I} + 2\boldsymbol{B}\frac{\partial w}{\partial I_1} - 2\boldsymbol{B}^{-1}\frac{\partial w}{\partial I_2} \tag{7.5}$$

for an incompressible hyperelastic solid and

$$\boldsymbol{\sigma} = \frac{2}{\sqrt{I_3}}\left\{\left(I_2\frac{\partial \hat{w}}{\partial I_2} + I_3\frac{\partial \hat{w}}{\partial I_3}\right)\mathrm{I} + \frac{\partial \hat{w}}{\partial I_1}\boldsymbol{B} - I_3\frac{\partial \hat{w}}{\partial I_2}\boldsymbol{B}^{-1}\right\} \tag{7.6}$$

for a compressible hyperelastic solid.

7.3 Simple Shear of a Rectangular Block

We consider a block that is rectangular in the undeformed reference configuration with three sides coinciding with the coordinate axes as indicated in Figure 7.1.

The block is subjected to a simple shear deformation as indicated by the dashed lines in Figure 7.1 and the deformation field is homogeneous and given by

$$x_1 = X_1 + kX_2, \quad x_2 = X_2, \quad x_3 = X_3, \tag{7.7}$$

where the angle of shear is $\theta = \arctan k$. It follows that the matrices of components of $\boldsymbol{F}$, $\boldsymbol{B}$, and $\boldsymbol{B}^{-1}$ are

$$\begin{aligned} [\boldsymbol{F}] = [\mathrm{Grad}\boldsymbol{x}] &= \begin{bmatrix} 1 & k & 0 \\ 0 & 1 & 0 \\ 0 & 0 & 1 \end{bmatrix}, \\ [\boldsymbol{B}] = \left[\boldsymbol{F}\boldsymbol{F}^T\right] &= \begin{bmatrix} 1+k^2 & k & 0 \\ k & 1 & 0 \\ 0 & 0 & 1 \end{bmatrix}, \end{aligned} \tag{7.8}$$

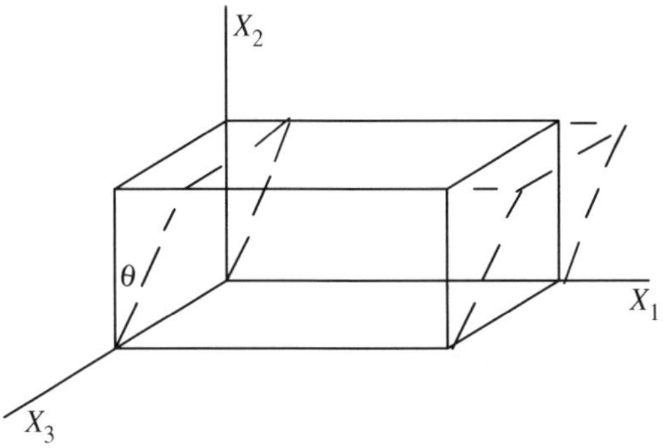

Figure 7.1. Simple shear of a rectangular block.

$$\left[\boldsymbol{B}^{-1}\right] = \begin{bmatrix} 1 & -k & 0 \\ -k & k^2+1 & 0 \\ 0 & 0 & 1 \end{bmatrix},$$

respectively.

Simple shear is a plane deformation and it is easily verified that $I_1 = I_2 = 3 + k^2$.

First we consider an incompressible solid and use equations (7.8) and constitutive equation (7.5) to obtain

$$\begin{aligned} \sigma_{11} &= -p + 2\frac{\partial w}{\partial I_1}\left(1+k^2\right) - 2\frac{\partial w}{\partial I_2}, \\ \sigma_{22} &= -p + 2\frac{\partial w}{\partial I_1} - 2\frac{\partial w}{\partial I_2}\left(1+k^2\right), \\ \sigma_{33} &= -p + 2\frac{\partial w}{\partial I_1} - 2\frac{\partial w}{\partial I_2}, \\ \sigma_{12} &= 2\left(\frac{\partial w}{\partial I_1} + \frac{\partial w}{\partial I_2}\right)k. \end{aligned} \tag{7.9}$$

It is clear that the normal stresses cannot be obtained from the deformation only, since the Lagrangian multiplier p depends on a stress boundary condition. If $\sigma_{33} = 0$, equations (7.9) give

$$\sigma_{11} = 2k^2\frac{\partial w}{\partial I_1}, \quad \sigma_{22} = -2k^2\frac{\partial w}{\partial I_2}, \quad \sigma_{12} = 2\left(\frac{\partial w}{\partial I_1} + \frac{\partial w}{\partial I_2}\right)k, \tag{7.10}$$

and for the Mooney-Rivlin strain energy function (7.1)

$$\sigma_{11} = \mu\gamma k^2, \quad \sigma_{22} = -\mu(1-\gamma)k^2, \quad \sigma_{12} = \mu k. \tag{7.11}$$

This analysis proceeds in the same way if σ_{11} or σ_{22} is taken as zero. It is evident from equations (7.10) and (7.11) that, in order to maintain the finite simple shear given by equation (7.7), normal stress components σ_{11} and σ_{22} are required for the Mooney-Rivlin and σ_{11} for the neo-hookean Hookean, in addition to σ_{12}. This is a second-order effect that is a form of the Poynting effect [8]. The normal components σ_{11} and σ_{22} are $O(k^2)$ for the Mooney-Rivlin solid, whereas there is a linear relation between σ_{12} and k. The relation between σ_{12} and k is nonlinear for materials for which $\partial w/\partial I_1$ and or $\partial w/\partial I_2$ are functions of k.

We now consider the compressible solid and use the constitutive equation (7.6). The stress components are now completely determined by the deformation equation (7.7) and for strain energy function (7.4) are obtained from equations (7.6) and (7.8) as

$$\sigma_{11} = \mu k^2, \quad \sigma_{22} = \sigma_{33} = 0, \quad \sigma_{12} = \mu k. \tag{7.12}$$

It is interesting to note that these components do not depend on γ and are the same as those for equation (7.11) with $\gamma = 1$, that is, for the neo-Hookean case with $\sigma_{33} = 0$.

7.4 Simple Tension

When an isotopic elastic solid is subjected to simple tension, due to a tensile stress $\sigma_{11} = \sigma$ and all other $\sigma_{ij} = 0$, the resulting homogeneous deformation is

$$x_1 = \lambda_1 X_1, \quad x_2 = \lambda_2 X_2, \quad x_3 = \lambda_3 X_3, \quad \lambda_2 = \lambda_3, \tag{7.13}$$

where $\lambda_1 > 1$ for simple tension and $0 < \lambda_1 < 1$ for simple compression. It follows from equation (7.13) that the matrix of components of the left Cauchy-Green strain tensor is given by

$$[\boldsymbol{B}] = \mathrm{diag}\left(\lambda_1^2, \lambda_2^2, \lambda_2^2\right), \tag{7.14}$$

and the principal invariants of $\boldsymbol{B}$ are

$$I_1 = \lambda_1^2 + 2\lambda_2^2, \quad I_2 = 2\lambda_1^2\lambda_2^2 + \lambda_2^4, \quad I_3 = \lambda_1^2\lambda_2^4. \tag{7.15}$$

First we consider a compressible solid. The following relations for the stresses are obtained from equations (7.6), (7.14), and (7.15),

$$\sigma_{11} = \frac{2\lambda_1}{\lambda_2^2}\left(\frac{\partial w}{\partial I_1} + 2\lambda_2^2\frac{\partial w}{\partial I_2}\right) + 2\lambda_1\lambda_2^2\frac{\partial w}{\partial I_3}, \tag{7.16}$$

$$\sigma_{22} = \sigma_{33} = \frac{2}{\lambda_1}\left\{\left(\frac{\partial w}{\partial I_1} + \left(\lambda_1^2 + \lambda_2^2\right)\frac{\partial w}{\partial I_2}\right)\right\} + 2\lambda_1\lambda_2^2\frac{\partial w}{\partial I_3} = 0. \tag{7.17}$$

The lateral stretch λ_2 is obtained from equation (7.17), in terms of the extension stretch λ_1; however, the form of $w(I_1, I_2, I_3)$ must be known in order to do this [9]. Substitution of λ_2 in equation (7.16) gives σ_{11} as a function of the stretch λ_1.

In order to illustrate this, strain energy function (7.4) is considered, so that

$$\frac{\partial w}{\partial I_1} = \frac{\mu}{2}, \quad \frac{\partial w}{\partial I_2} = 0, \quad \frac{\partial w}{\partial I_3} = \frac{\mu}{2}\left(\lambda_1^2\lambda_2^4\right)^{-(1-\nu)/(1-2\nu)},$$

which is substituted in equation (7.17). The resulting equation with given values of λ_1 and ν must be solved numerically for λ_2, which is then substituted in equation (7.16) to obtain σ_{11}.

Numerical results for $\lambda_1 = 1.25$ and $\nu = 0.49$, which is a realistic value for certain rubber-like materials [10], are

$$\lambda_2 = 0.89643, \quad \frac{\sigma_{11}}{\mu} = 0.75552.$$

The nominal stress for simple tension is given by $\Sigma_{11} = \sigma_{11}\lambda_2^2$.

The problem for an incompressible solid is simpler since the incompressibility condition $I_3 = 1$ gives

$$\lambda_2 = \lambda_3 = \lambda_1^{-1/2}.$$

PROBLEM. Show that, for simple tension of an incompressible solid,

$$\sigma_{11} = 2\left(\lambda_1^2 - \lambda_1^{-1}\right)\left(\frac{\partial w}{\partial I_1} + \frac{1}{\lambda_1}\frac{\partial w}{\partial I_2}\right),$$

and in particular for a Mooney-Rivlin solid,

$$\sigma_{11} = \mu\left(\lambda_1^2 - \lambda_1^{-1}\right)\left\{\gamma + \frac{(1-\gamma)}{\lambda_1}\right\}$$

and

$$\frac{N}{\pi A^2 \mu} = \left(\lambda_1 - \lambda_1^{-2}\right)\left(\gamma + \frac{(1-\gamma)}{\lambda_1}\right),$$

where N is the axial force and A is the radius in the undeformed reference configuration.

For simple tension of the neo-Hookean solid with $\lambda_1 = 1.25$ it follows from the results of this problem that

$$\lambda_2 = 0.89443, \quad \frac{\sigma_{11}}{\mu} = 0.7625,$$

which differ negligibly from the corresponding values for the compressible case already considered. This indicates that, for static problems of rubber-like materials, the incompressible model is justified unless a hydrostatic compressive stress of the order of magnitude of the bulk modulus is applied.

The Mooney-Rivlin strain energy function and its special case, the neo-Hookean, give results, for simple tension, with stretch less than about two that are in good agreement with experiment. For stretches greater than about three experimentally obtained nominal stress-stretch curves for most rubbers are S shaped with an inflexion point at about $\lambda = 3$. This cannot be predicted by the Mooney-Rivlin strain energy function. Ogden [5] has proposed a three-term strain energy function, that is capable of modeling nominal stress vs. stretch curves for large stretches in simple tension,

$$w(\lambda_1, \lambda_2, \lambda_3) = \sum_{i=1}^{3} \frac{\mu_i}{\alpha_i}(\lambda_1^{\alpha_i} + \lambda_2^{\alpha_i} + \lambda_3^{\alpha_i} - 3), \quad \lambda_1\lambda_2\lambda_3 = 1, \tag{7.18}$$

where

$$\sum_{i=1}^{3} \mu_i \alpha_i = 2\mu,$$

and the α_i and μ_i are chosen to fit experimental data. For simple tension of an incompressible solid the axial stretch $\lambda_1 = \lambda$ and $\lambda_2 = \lambda_3 = 1/\sqrt{\lambda}$ and it follows from equation (7.18) that the axial tensile nominal stress is given by

$$\Sigma = \frac{dw(\lambda)}{d\lambda} = \sum_{i=1}^{3} \mu_i \left(\lambda^{\alpha_i - 1} - \lambda^{-\alpha_i/2-1}\right). \tag{7.19}$$

Ogden [5] has shown that equation (7.19) gives a close agreement with experimental data for several rubbers for stretches up to seven when

$$\begin{aligned} &\mu_1/\mu = 1.491, \quad &&\mu_2/\mu = 0.003, \quad &&\mu_3/\mu = -\,0.0237, \\ &\alpha_1 = 1.3 &&\alpha_2 = 5.0 &&\alpha_3 = -2. \end{aligned} \tag{7.20}$$

Stress-stretch relations, obtained from equations (7.19) and (7.20), and the Mooney Mooney-Rivlin strain energy function with $\gamma = 1$ (neo-Hookean) and $\gamma = 0.6$, are shown graphically in Figure 7.2.

7.5 Extension and Torsion of an Incompressible Cylindrical Bar

In this section we consider the extension and torsion of an incompressible right circular cylinder. Cylindrical polar coordinates are used and the analysis is based on the physical components of the stress and deformation gradient tensors. The deformation is a controllable deformation.

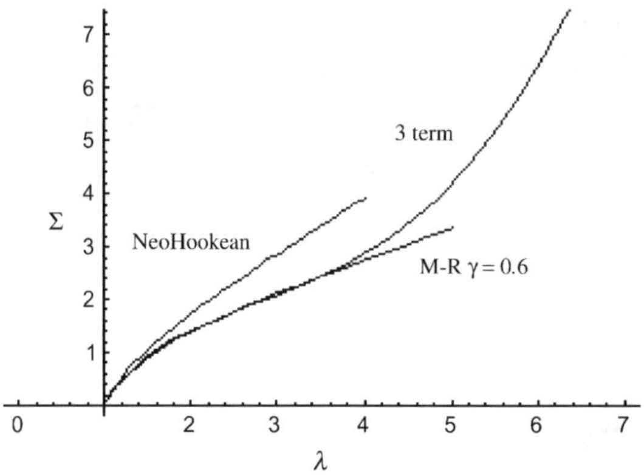

Figure 7.2. Nominal stress-stretch curves for simple tension.

The z axis of the cylindrical polar system is chosen to coincide with the axis of the cylinder and the origin is taken at one end of the cylinder, and the cylindrical polar coordinates of a particle in the spatial and undeformed reference configurations are denoted by (r, θ, z) and (R, Θ, Z), respectively, and the corresponding radii of the cylinder are $r = a$ and $R = A$. Deformation of the cylinder is a pure torsion with plane cross sections remaining plane followed by a homogeneous extension or vice versa, and is given by,

$$r = R\lambda^{-1/2}, \quad \theta = \Theta + DZ, \quad z = \lambda Z, \tag{7.21}$$

where D is the angle of twist per unit length of the reference configuration. It is reasonable to expect that a stress system equivalent to a torque T and axial force N is required to maintain the deformation (7.21).

The physical components of the deformation gradient tensor, left Cauchy-Green strain tensor $\boldsymbol{B}$, and $\boldsymbol{B}^{-1}$ are

$$\begin{aligned}
[\boldsymbol{F}] &= \begin{bmatrix} \lambda^{-1/2} & 0 & 0 \\ 0 & \lambda^{-1/2} & rD \\ 0 & 0 & \lambda \end{bmatrix}, \\
[\boldsymbol{B}] = \left[\boldsymbol{F}\boldsymbol{F}^{T}\right] &= \begin{bmatrix} \lambda^{-1} & 0 & 0 \\ 0 & \lambda^{-1} + r^2D^2 & \lambda rD \\ 0 & \lambda rD & \lambda^2 \end{bmatrix},
\end{aligned} \tag{7.22}$$

$$\left[\boldsymbol{B}^{-1}\right] = \begin{bmatrix} \lambda & 0 & 0 \\ 0 & \lambda & -rD \\ 0 & -rD & \left(\lambda^{-2} + \lambda^{-1} r^2 D^2\right) \end{bmatrix},$$

respectively.

A solution is now obtained and particularized for the Mooney-Rivlin material. The procedure given can be adapted to consider any incompressible hyperelastic material, although more complication results when $\partial w/\partial I_1$ and/or $\partial w/\partial I_2$ are not constant.

Expressions that involve the Lagrangian multiplier p are obtained from equations (7.1), (7.5), and (7.22) for the nonzero components of the Cauchy stress,

$$\sigma_r = -p + 2\left\{\lambda^{-1}\frac{\partial w}{\partial I_1} - \lambda\frac{\partial w}{\partial I_2}\right\}, \tag{7.23}$$

$$\sigma_\theta = -p + 2\left\{\left(\lambda^{-1} + r^2 D^2\right)\frac{\partial w}{\partial I_1} - \lambda\frac{\partial w}{\partial I_2}\right\}, \tag{7.24}$$

$$\sigma_z = -p + 2\left(\lambda^2\frac{\partial w}{\partial I_1} - \left(\lambda^{-2} + \lambda^{-1} r^2 D^2\right)\frac{\partial w}{\partial I_2}\right), \tag{7.25}$$

$$\sigma_{\theta z} = rD\left\{\lambda\frac{\partial w}{\partial I_1} + \frac{\partial w}{\partial I_2}\right\}. \tag{7.26}$$

The only nonzero trivial equilibrium equation is

$$\frac{d\sigma_r}{dr} + \frac{\sigma_r - \sigma_\theta}{r} = 0. \tag{7.27}$$

Substitution of equations (7.23) and (7.24) in equation (7.27) gives

$$\frac{d\sigma_r}{dr} = -2rD^2\frac{\partial w}{\partial I_1},$$

which, along with the boundary condition $\sigma_r(a) = 0$, gives

$$\sigma_r = -2 \int_r^a rD^2 \frac{\partial w}{\partial I_1} dr. \tag{7.28}$$

then p is obtained from equations (7.23) and (7.28) and this enables it to be eliminated from equations (7.24) and (7.25).

It is clear from the analysis so far that equation (7.21) is a controllable deformation. If the Mooney-Rivlin strain energy function is considered, the solution is relatively simple since

$$\partial w/\partial I_1 = \mu\gamma/2 \text{ and } \partial w/\partial I_2 = \mu(1-\gamma)/2$$

are constants and it then follows from equation (7.23) that $\partial p/\partial r = -\partial\sigma_r/\partial r$. The following results are for the Mooney-Rivlin strain energy function:

$$\frac{p}{\mu} = \frac{\gamma D^2}{2}\left(a^2 - r^2\right) + \gamma\lambda^{-1} - (1-\gamma)\lambda,$$

$$\frac{\sigma_r}{\mu} = -\frac{\gamma D^2}{2}\left(a^2 - r^2\right), \tag{7.29}$$

$$\frac{\sigma_\theta}{\mu} = -\frac{\gamma D^2}{2}\left(a^2 - 3r^2\right), \tag{7.30}$$

$$\begin{aligned}\frac{\sigma_z}{\mu} = {} & D^2r^2\left\{\frac{\gamma}{2} - (1-\gamma)\lambda^{-1}\right\} - \frac{\gamma D^2 a^2}{2} \\ & + (1-\gamma)\left(\lambda - \lambda^{-2}\right) + \gamma\left(\lambda^2 - \lambda^{-1}\right),\end{aligned} \tag{7.31}$$

$$\frac{\sigma_{\theta z}}{\mu} = \{\gamma(\lambda - 1) + 1\}rD. \tag{7.32}$$

The torque T required is

$$T = 2\pi \int_0^a r^2 \sigma_{\theta z} dr,$$

and equation (7.32) and the incompressibility condition, $a^2 = \lambda^{-1} A^2$, give the nondimensional form,

$$\frac{T}{\pi \mu A^3} = \frac{\bar{D}}{2\lambda^4}[\gamma(\lambda - 1) + 1], \tag{7.33}$$

where $\bar{D} = DA$.

Similarly, the axial force required is given by

$$N = 2\pi \int_0^a r \sigma_z dr,$$

and with the use of equation (7.28) this gives the nondimensional form

$$\frac{N}{\pi \mu A^2} = \frac{\bar{D}^2}{2}\left\{\gamma\left(\lambda^{-3} - \frac{\lambda^{-2}}{2}\right) - \lambda^{-3}\right\} + (1-\gamma)\left(1 - \lambda^{-3}\right) + \gamma\left(\lambda - \lambda^{-2}\right). \tag{7.34}$$

Equation (7.33) shows that, for the Mooney-Rivlin material, the relation between torque and angle of twist per unit length is linear for a fixed value of λ. Equation (7.34) shows that if the length of the bar is constrained with $\lambda = 1$ an axial force given by

$$\frac{N}{\pi \mu A^2} = -\frac{\bar{D}^2 \gamma}{4}$$

is required, and if the axial force is zero, the length increases. This is a second-order effect, known as the Poynting effect [8], which does not exist in linear elasticity theory.

7.6 Spherically Symmetric Expansion of a Thick-Walled Shell

A thick-walled incompressible hyperelastic spherical shell is considered and a solution is obtained for the Mooney-Rivlin strain energy function. This solution is similar to a more general solution given by Green and Zerna [3]. A spherical polar coordinate system with origin at the center of the shell is used and the deformation is given by

$$r = f(R), \quad \theta = \Theta, \quad \phi = \Phi \tag{7.35}$$

where r, θ, and ϕ are the spherical polar coordinates as defined in Appendix 1.

As usual in this text lower- (upper-) case letters refer to the spatial (reference) configurations. The inner radii in the undeformed natural reference configuration and spatial configuration are A and a, respectively, and the corresponding outer radii are B and b.

Nonzero components of the deformation gradient tensor are identical to the principal stretches $(\lambda_r, \lambda_\theta, \lambda_\phi)$ where it follows from incompressibility that $\lambda_r\lambda_\theta\lambda_\phi = 1$. It may be deduced from equation (7.35) that $\lambda_\theta = \lambda_\phi = r/R$ and from the incompressibility condition that $\lambda_r = (R/r)^2$. The components of $\boldsymbol{B}$ and $\boldsymbol{B}^{-1}$ are then given by

$$[\boldsymbol{B}] = \text{diag}\left[Q^4, Q^{-2}, Q^{-2}\right], \quad \left[\boldsymbol{B}^{-1}\right] = \text{diag}\left[Q^{-4}, Q^2, Q^2\right], \tag{7.36}$$

where $Q = R/r$. Expressions, involving the Lagrangian multiplier p, for the stresses,

$$\sigma_r = -p + 2\left(Q^4\frac{\partial w}{\partial I_1} - Q^{-4}\frac{\partial w}{\partial I_2}\right), \tag{7.37}$$

$$\sigma_\theta = \sigma_\phi = -p + 2\left(Q^{-2}\frac{\partial w}{\partial I_1} - Q^2\frac{\partial w}{\partial I_2}\right), \tag{7.38}$$

are obtained from equations (7.5) and (7.36). The nontrivial equilibrium equation is

$$\frac{d\sigma_r}{dr} + \frac{2(\sigma_r - \sigma_\theta)}{r} = 0, \tag{7.39}$$

and it follows from the incompressibility condition $r^3 - a^3 = R^3 - A^3$ that

$$\frac{dQ}{dr} = \frac{1}{r}\left(Q^{-2} - Q\right).$$

Substituting equations (7.37), (7.38), and $\frac{d\sigma_r}{dr} = \frac{d\sigma_r}{dQ}\frac{dQ}{dr}$ in equation (7.39) gives

$$\frac{d\sigma_r}{dQ} = -4\left\{\left(Q^4 - Q^{-2}\right)\frac{\partial w}{\partial I_1} - \left(Q^{-4} - Q^2\right)\frac{\partial w}{\partial I_2}\right\} \Big/ \left(Q^{-2} - Q\right) \tag{7.40}$$

Equation (7.40) can be integrated and the constant of integration determined by a boundary condition. This integration is simplified if $\partial w/\partial I_1$ and $\partial w/\partial I_2$ are constants as in the Mooney-Rivlin strain energy function (7.1), then equation (7.40) becomes

$$\frac{d\sigma_r}{dQ} = -2\mu\left\{\left(Q^4 - Q^{-2}\right)\gamma - \left(Q^{-4} - Q^2\right)(1-\gamma)\right\} \Big/ \left(Q^{-2} - Q\right). \tag{7.41}$$

A solution is now obtained for the Mooney-Rivlin strain energy function. The integral of equation (7.41) is

$$\frac{\sigma_r}{\mu} = \left(\frac{Q^4}{2} + 2Q\right)\gamma + 2(1-\gamma)\left(\frac{Q^2}{2} - Q^{-1}\right) + K, \tag{7.42}$$

where K is a constant of integration that is evaluated from the boundary conditions

$$\sigma_r(Q_i) = -P_i \quad \text{and} \quad \sigma_r(Q_o) = -P_0, \tag{7.43}$$

where $Q_i = A/a$ and $Q_o = B/b$ and P_i and P_0 are the internal and external pressures, respectively. Since the shell is assumed to be incompressible the deformation due to equation (7.43) is the same as that due to

$$\sigma_r(Q_i) = -\hat{P}_i \text{ and } \sigma_r(Q_o) = 0, \tag{7.44}$$

where $\hat{P}_i = P_i - P_o$. Consequently, the solution is obtained for boundary conditions (7.44). From equations (7.42) and $(7.44)_1$

$$\left(\frac{Q_i{}^4}{2} + 2Q_i\right)\gamma + 2(1-\gamma)\left(\frac{Q_i^2}{2} - Q_i^{-1}\right) + K = -\frac{\hat{P}_i}{\mu}, \tag{7.45}$$

Eliminating K from equations (7.42) and (7.45) gives

$$\begin{aligned}\frac{\sigma_r}{\mu} = {} & \gamma\left\{\frac{\left(Q^4 - Q_i^4\right)}{2} + 2(Q - Q_i)\right\} \\ & + 2(1-\gamma)\left\{\left(\frac{\left(Q^2 - Q_i^2\right)}{2} - \left(Q^{-1} - Q_i^{-1}\right)\right)\right\} - \frac{\hat{P}_i}{\mu},\end{aligned} \tag{7.46}$$

and eliminating p from equations (7.37) and (7.38) to obtain $\sigma_\theta - \sigma_r$ and adding σ_r from equation (7.46) gives

$$\begin{aligned}\frac{\sigma_\theta}{\mu} = \frac{\sigma_\phi}{\mu} = {} & \gamma\left\{Q^{-2} + 2(Q - Q_i) - \frac{\left(Q^4 - Q_i^4\right)}{2}\right\} \\ & + 2(1-\gamma)\left\{\frac{\left(Q^2 - Q_i^2\right)}{2} - \left(Q^{-1} - Q_i^{-1}\right)\right\} - \frac{\hat{P}_i}{\mu}.\end{aligned} \tag{7.47}$$

It is desirable to have a relation between $Q_i = A/a$ and $\hat{P}_i$ for a given value of $\eta = A/B$. It follows from equations $(7.44)_2$ and (7.46) that

$$\begin{aligned}\frac{\hat{P}_i}{\mu} = {} & \gamma\left\{\frac{\left(Q_o^4 - Q_i^4\right)}{2} + 2(Q_0 - Q_i)\right\} \\ & + 2(1-\gamma)\left\{\frac{\left(Q_o^2 - Q_i^2\right)}{2} - \left(Q_o^{-1} - Q_i^{-1}\right)\right\}.\end{aligned} \tag{7.48}$$

The incompressibility condition $B^3 - A^3 = b^3 - a^3$ gives, after some manipulation, the relation

$$Q_0 = \frac{Q_i}{\left\{\eta^3 + \left(1 - \eta^3\right)Q_i^3\right\}^{1/3}}. \tag{7.49}$$

Elimination of Q_o from equations (7.48) and (7.49) gives the non dimensional relation between $\hat{P}_i/\mu$ and $a/A = Q_1^{-1}$ for a given value of η. This relation is shown graphically in Figure 7.3 for the Neo-Hookean model ($\gamma = 1$) and $\eta = 0.5$.

Instability of the shell occurs when

$$\frac{d(\hat{P}_i/\mu)}{d(a/A)} = 0.$$

For the above data the onset of instability occurs for

$$\hat{P}_i/\mu = 0.8164 \text{ and } a/A = 1.828.$$

A thin-walled spherical shell is a limiting case of the problem just considered. For a thin-walled shell, the wall thickness in the undeformed reference configuration is $T = B - A = A(\eta^{-1} - 1) << A$ and $\sigma_\theta = \sigma_\phi >> \sigma_r(a) = \hat{P}_i$. The variation $\sigma_\theta = \sigma_\phi$ is assumed to be constant across

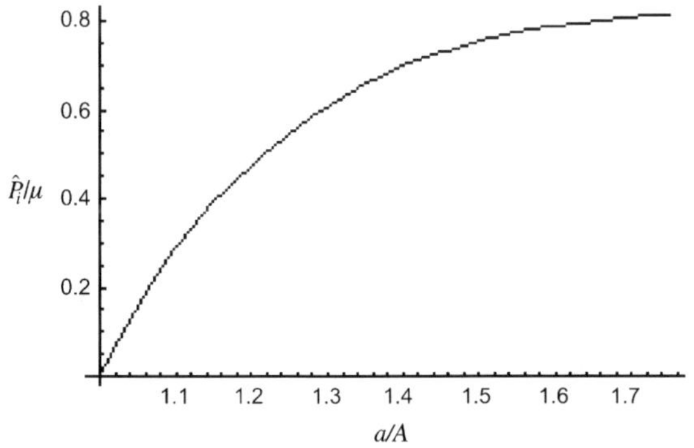

Figure 7.3. Nondimensional internal pressure-internal radius relation for thick-walled Neo-Hookean shell with $\eta = 0.5$.

the wall thickness and $\sigma_r(r)$ are neglected, and the elementary approximate relation

$$\hat{P}_i \simeq \frac{2\sigma_\theta t}{a} \tag{7.50}$$

is obtained, where $t = b - a$ is the wall thickness in the spatial configuration. It also follows from the approximations made that

$$Q^{-1} \simeq \frac{a}{A},$$

and the thin-walled shell is approximately in a state of equibiaxial stress so that for the neo-Hookean solid

$$w = \frac{\mu}{2}(2\lambda^2 + \lambda^{-4}),$$

where, for the thin-walled shell problem, $\lambda \simeq \sqrt{a/A}$. Then, it may be deduced that

$$\sigma_\theta = \frac{1}{2}\frac{dw(\lambda)}{d\lambda},$$

which gives

$$\frac{\sigma_\theta}{\mu} \simeq \left\{\left(\frac{a}{A}\right)^2 - \left(\frac{a}{A}\right)^{-4}\right\}. \tag{7.51}$$

It follows from the incompressibility condition that $t/T \simeq (a/A)^{-2}$ and

$$\frac{t}{a} = \frac{t}{T}\frac{T}{a} = \left(\frac{a}{A}\right)^{-3}\left(\eta^{-1} - 1\right). \tag{7.52}$$

Substituting equations (7.51) and (7.52) into equation (7.50) gives

$$\frac{\hat{P}_i}{\mu} = 2\left(\eta^{-1} - 1\right)\left\{\left(\frac{a}{A}\right)^{-1} - \left(\frac{a}{A}\right)^{-7}\right\}. \tag{7.53}$$

In Figure 7.4, the relation between $\hat{P}_i / \mu$ and a/A is shown graphically for the neo-Hookean model and $\eta = 0.975$. The dashed curve is obtained from equation (7.53) and the solid curve is obtained from equations (7.48) and (7.49). It is evident that, for thin-walled spherical shell, the approximate theory is in close agreement with the theory for the thick-walled shell.

7.7 Eversion of a Cylindrical Tube

Eversion of a hyperelastic circular cylindrical tube is the turning inside out. This is an example of a nonunique solution for the deformation of a hyperelastic solid subjected to prescribed surface tractions, in this case zero surface tractions. A formal solution of this eversion problem has been given by Rivlin [11], and later by Wang and Truesdell [12]. The solution given in [12] involves two complicated nonlinear algebraic simultaneous equations. According to [12] it is not known whether or not these equations have an admissible solution except for the Mooney-Rivlin strain energy function or its special case, the neo-Hookean. The solution given here for the neo-Hookean solid is essentially that given in [12] and results for a particular example are given.

The inside and outside radii of the tube in the (reference, spatial) configuration are denoted by (A, a) and (B, b), respectively, and the

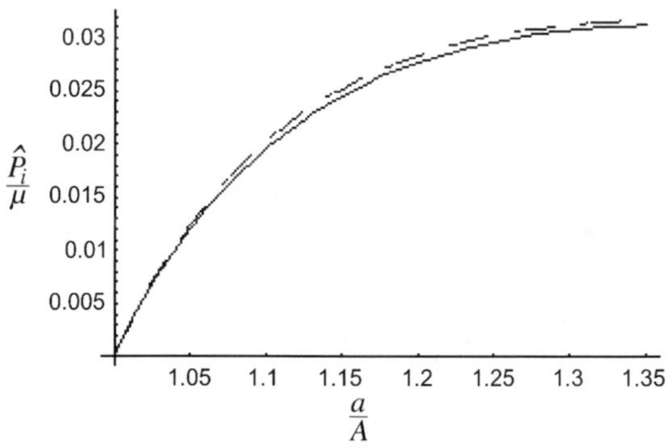

Figure 7.4. Comparison of results for $\eta = 0.975$ obtained from thick-walled theory and approximate thin-walled theory.

eversion process is shown schematically in Figure 7.5 for length (L, l) of a infinite tube in the (reference, spatial) configuration. For a finite tube it is assumed that it is sufficiently long that warping of the ends can be neglected and the resultant axial force is zero. It follows that the deformation field is given by

$$r = f(R), \quad \theta = \Theta, \quad z = -\lambda Z, \tag{7.54}$$

where $\lambda = l/L$, and the physical components of the deformation gradient tensor are given by

$$[\boldsymbol{F}] = \text{diag}\,[dr/dR,\, r/R, -\lambda]. \tag{7.55}$$

Since incompressibility is assumed, $\det[\boldsymbol{F}] = 1$, and it follows from equation (7.55) that

$$\frac{dr}{dR} = -\frac{R}{\lambda r}. \tag{7.56}$$

Integration then gives

$$r = \sqrt{\alpha R^2 + \beta}, \tag{7.57}$$

where $\alpha = -\lambda^{-1} < 0$ and β is a constant of integration.

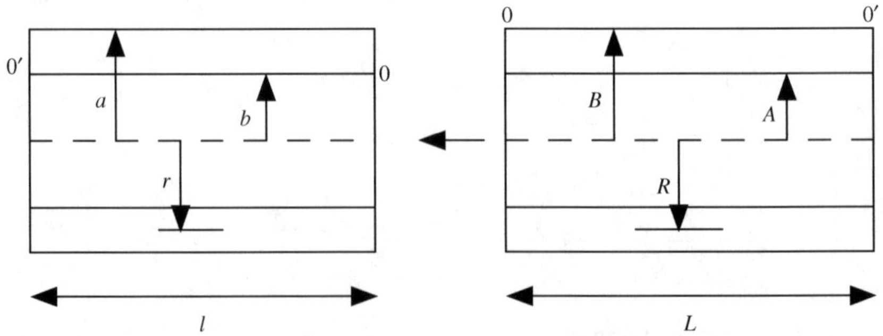

Figure 7.5. Eversion of a cylindrical tube.

It is useful to introduce nondimensional quantities by dividing quantities with dimension of length by A and stresses by μ, so that

$$(\bar{r}, \bar{R}) = \frac{(r, R)}{A}, \quad \bar{\beta} = \frac{\beta}{A^2}, \quad (\bar{\sigma}_r, \bar{\sigma}_\theta, \bar{\sigma}_z) = \frac{(\sigma_r, \sigma_\theta, \sigma_z)}{\mu},$$
$$\bar{A} = 1, \quad \bar{B} = \frac{B}{A} = \eta.$$

Henceforth nondimensional quantities are used but the superposed bars are omitted for convenience. Components

$$[\boldsymbol{B}] = \operatorname{diag}\left[\alpha^2 \frac{R^2}{r^2}, \frac{r^2}{R^2}, \alpha^{-2}\right] \tag{7.58}$$

of the left Cauchy-Green tensor, $\boldsymbol{B} = \boldsymbol{F}\boldsymbol{F}^T$, are obtained from equations (7.55) and (7.56). The nonzero components of Cauchy stress for the neo-Hookean solid are

$$\sigma_r = -p + \left(\frac{\alpha R}{r}\right)^2, \tag{7.59}$$

$$\sigma_\theta = -p + \left(\frac{r}{R}\right)^2. \tag{7.60}$$

$$\sigma_z = -p + \alpha^{-2}. \tag{7.61}$$

Substitution of equations (7.59) and (7.60) in the nontrivial equilibrium equation

$$\frac{\partial \sigma_r}{\partial r} + \frac{\sigma_r - \sigma_\theta}{r} = 0$$

gives

$$\sigma_r = -\int \left(\frac{\alpha^2 R^2}{r^2} - \frac{r^2}{R^2}\right) \frac{dr}{r}. \tag{7.62}$$

Since $\sigma_r(a) = \sigma_r(b) = 0$ the integral (7.62) with limits of integration a and b is zero, that is,

$$\int_a^b \left(\frac{\alpha^2 R^2}{r^2} - \frac{r^2}{R^2} \right) \frac{dr}{r} = 0. \tag{7.63}$$

Substitution of equation (7.57) in equation (7.63) gives the non dimensional relation, referred to the reference configuration,

$$\int_1^\eta \left[\frac{\alpha^2 R^2}{\left(\alpha R^2 + \beta\right)^2} - \frac{1}{R^2} \right] R dR = 0. \tag{7.64}$$

The everted tube is assumed to have no resultant axial force applied so that

$$2\pi \int_a^b r \sigma_z dr = 0. \tag{7.65}$$

Using equations (7.57) and (7.59) to (7.61), it may be deduced from equation (7.65) that

$$\int_1^\eta \left[\frac{\alpha^2 R^2}{\alpha R^2 + \beta} - \frac{2}{\alpha^2} + \frac{\alpha R^2 + \beta}{R^2} \right] R dR = 0. \tag{7.66}$$

Equations (7.64) and (7.66) can be solved simultaneously to obtain α and β. It appears to be very difficult to determine closed-form expressions for α and β and a numerical determination for a given value of η is desirable.

A numerical example is now considered for the neo-Hookean solid with $\eta = 1.15$. Then, a numerical solution of equations (7.64) and (7.66) obtained using Mathematica gives

$$\alpha = -\ 0.995729, \quad \beta = 2.32722.$$

Since $\lambda = -\alpha^{-1} > 1$ the tube lengthens when it is inverted. The non dimensional radii in the deformed configuration

$$a = 1.1539 \quad \text{and} \quad b = 1.00517$$

are then obtained from equation (7.57). It is easily verified that the isochoric condition

$$\left(\eta^2 - 1\right) = -\alpha^{-1}\left(a^2 - b^2\right)$$

is satisfied.

The Mooney-Rivlin solid can be considered with some additional complication.

7.8 Pure Bending of an Hyperelastic Plate

Pure bending of an incompressible isotropic hyperelastic plate is a further example of a controllable deformation. Cuboid, rather than plate, is the term used by Rivlin [11], who appears to be the first to have considered this problem that has also been considered by Ogden [5].

The natural reference configuration of the plate is bounded by the planes

$$X_1 = S_1, \quad X_1 = S_2, \quad X_2 = \pm L, \quad X_3 = \pm W/2$$

and

$$S_2 - S_1 = H.$$

It is assumed that the plate is subjected to pure bending by normal stresses, statically equivalent to couples, on faces $X_2 = \pm L$, and it is bent into a sector of a hollow circular cylinder with inner and outer radii a and b, respectively, as indicated in Figure 7.6. The bent deformed configuration

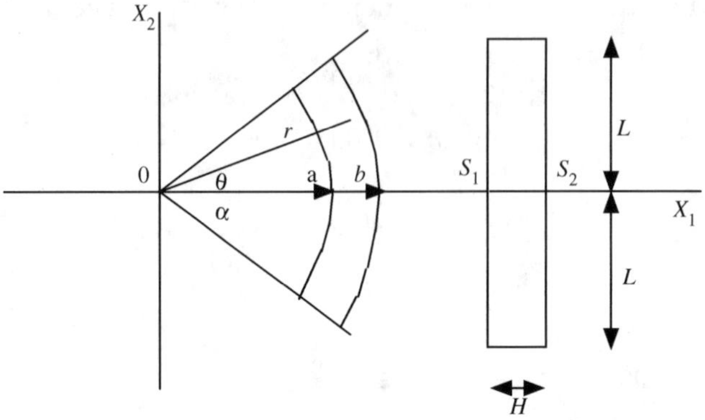

Figure 7.6. Finite deformation pure bending of hyperelastic plate.

is described by the cylindrical polar coordinates (r, θ, z) and is bounded by surfaces

$$r = a,\ r = b,\ \theta = \alpha,\ \theta = -\alpha,\ z = \mathrm{w},\ z = -\mathrm{w}.$$

For the pure bending, planes normal to the X_2 axis in the undeformed reference configuration are plane in the deformed configuration and are in meridian planes containing the X_3 axis. Also the stretch, λ, in the X_3 direction is uniform, so that $z = \lambda X_3$, $\lambda = \mathrm{w}/\mathrm{W}$, which implies that anticlastic bending is constrained.

A material plane $X_3 = K$, where $S_1 \leqslant K \leqslant S_2$ is a constant, becomes a sector of a cylinder of radius r subtending the angle 2α at the center. It then follows that the circumferential stretch in the spatial configuration is given by

$$\lambda_\theta = \frac{r\alpha}{L}. \tag{7.67}$$

Since the solid is incompressible the radial stretch is given by

$$\lambda_r = \frac{L}{r\alpha\lambda}. \tag{7.68}$$

And since the r, θ, and z directions are principal directions, the physical components of the left Cauchy strain tensor are given by

$$[\boldsymbol{B}] = \operatorname{diag}\left[\frac{L^2}{\lambda^2 r^2 \alpha^2}, \frac{r^2\alpha^2}{L^2}, \lambda^2\right]. \tag{7.69}$$

A further relation

$$\frac{L}{\alpha} = \frac{\lambda\left(b^2 - a^2\right)}{2H} \tag{7.70}$$

is obtained from the incompressibility condition. Relations (7.67) to (7.70) are valid for any isotropic incompressible strain energy function.

A solution is now outlined for the neo-Hookean material and solutions can readily be obtained for other isotropic hyperelastic solids, especially the Mooney-Rivlin, with some additional complication.

The nonzero physical components of Cauchy stress, obtained from equations (7.5) and (7.68) for the neo-Hookean solid, are given by

$$\frac{\sigma_r}{\mu} = -\frac{p}{\mu} + \frac{L^2}{\lambda^2 r^2 \alpha^2}, \tag{7.71}$$

$$\frac{\sigma_\theta}{\mu} = -\frac{p}{\mu} + \frac{r^2\alpha^2}{L^2}, \tag{7.72}$$

$$\frac{\sigma_z}{\mu} = -\frac{p}{\mu} + \lambda^2, \tag{7.73}$$

to within the Lagrangian multiplier p.

Substitution of equations (7.71) and (7.72) in the equilibrium equation (7.27)

$$\frac{d\sigma_r}{dr} + \frac{(\sigma_r - \sigma_\theta)}{r} = 0$$

gives

$$\frac{1}{\mu}\frac{d\sigma_r}{dr} = -\left[\frac{L^2}{\lambda^2 r^2 \alpha^2} - \frac{r^2\alpha^2}{L^2}\right]\Big/ r,$$

so that, upon integration,

$$\frac{\sigma_r}{\mu} = \frac{1}{2}\left[\frac{L^2}{\lambda^2 r^2 \alpha^2} + \frac{r^2\alpha^2}{L^2}\right] + c_1, \tag{7.74}$$

where c_1 is a constant of integration. It then follows from the boundary conditions,

$$\sigma_r(a) = 0, \quad \sigma_r(b) = 0,$$

that

$$c_1 = -\frac{1}{2}\left(\frac{L^2}{\lambda^2 b^2 \alpha^2} + \frac{b^2\alpha^2}{L^2}\right) = -\frac{1}{2}\left(\frac{L^2}{\lambda^2 a^2 \alpha^2} + \frac{a^2\alpha^2}{L^2}\right). \tag{7.75}$$

The stress σ_r is obtained by eliminating c_1 from equations (7.74) and (7.75). Solving the second equation of equation (8.75) for L^2/α^2 gives

$$\frac{L^2}{\alpha^2} = \lambda ab. \tag{7.76}$$

If λ and either α or one of the radii of the cylindrical surfaces are given, equations (7.70) and (7.76) determine the deformation.

It follows from equations (7.71) and (7.74) that

$$-\frac{p}{\mu} = \frac{1}{2}\left(\frac{r^2\alpha^2}{L^2} - \frac{L^2}{\lambda^2 r^2 \alpha^2}\right) + c_1, \tag{7.77}$$

and then from equations (7.72) and (7.73) that

$$\frac{\sigma_\theta}{\mu} = \frac{1}{2}\left(3\frac{r^2\alpha^2}{L^2} - \frac{L^2}{\lambda^2 r^2 \alpha^2}\right) + c_1, \tag{7.78}$$

$$\frac{\sigma_z}{\mu} = \lambda^2 + \frac{1}{2}\left(\frac{r^2\alpha^2}{L^2} - \frac{L^2}{\lambda^2 r^2 \alpha^2}\right) + c_1, \tag{7.79}$$

where c_1 is given by equation (7.75).

The resultant force acting on the surfaces $z = \pm$w is given by

$$F_z = 2\alpha \int_a^b r\sigma_z dr, \tag{7.80}$$

and the distribution of σ_θ acting on surfaces $\theta = \pm\alpha$ is statically equivalent to the couple

$$M_z = 2w \int_a^b r\sigma_\theta dr. \tag{7.81}$$

The length of a material line parallel to the X_2 axis in the reference configuration is $2L$, and in the spatial configuration, this line is bent into an arc of a circle of radius r and length $r\alpha$. If $\lambda = 1$ and the length of the line are unchanged due to the deformation, it follows from equation (7.76) that $r = r_0$ where $r_0 = (ab)^{1/2}$, and then from equation (7.68) that the principal stretches on surface $r = r_0$ are $(1, 1, 1)$. Consequently, on the surface $r = r_0$ the stress is isotropic and from equations (7.74) to (7.76), (7.78) and (7.79) is given by

$$\sigma_r(r_0)/\mu = \sigma_\theta(r_0)/\mu = \sigma_z(r_0)/\mu = 1 + c_1. \tag{7.82}$$

Numerical examples are now given to illustrate the solution for $\lambda = 1$. Nondimensional quantities are used and are denoted by a superposed bar.

Quantities with dimension of length and stress are nondimensionalized by dividing by L and μ, respectively, so that

$$\bar{r} = r/L,\ \boldsymbol{\sigma} = \bar{\boldsymbol{\sigma}}/\mu,\quad \bar{H}=H/L,\quad \bar{W}=W/L,\quad \bar{F} = F\big/\mu L^2,\quad \bar{M} = M\big/\mu L^3.$$

The nondimensional forms of equations (7.70) and (7.76), with $\lambda = 1$, are

$$\frac{\alpha\left(\bar{b}^2 - \bar{a}^2\right)}{2\bar{H}} = 1 \quad \text{and} \quad \alpha^2\bar{a}\bar{b} = 1, \tag{7.83}$$

respectively, which are solved simultaneously to obtain $\bar{a}$ and $\bar{b}$. The nondimensional resultant force,

$$\bar{F}_z = 2\alpha \int_{\bar{a}}^{\bar{b}} \bar{r}\left\{1 + \frac{1}{2}\left\{(\alpha\bar{r})^2 - (\alpha\bar{r})^{-2}\right\} + c_1\right\} d\bar{r}, \tag{7.84}$$

acting on surfaces $z = \pm w$ is obtained from equations (7.79) and (7.80) and the distribution of $\bar{\sigma}_\theta$ acting on surfaces $\theta = \pm\alpha$ is statically equivalent to the nondimensional couple per unit width of the reference configuration,

$$\overline{M_z/2W} = \int_{\bar{a}}^{\bar{b}} \bar{r}\left\{\frac{1}{2}\left(3(\alpha\bar{r})^2 - (\alpha\bar{r})^{-2}\right) + c_1\right\} d\bar{r}, \tag{7.85}$$

obtained from equations (7.78) and (7.81). The stress distribution $\bar{\sigma}_z(\bar{r})$ at $z = \pm w$ is statically equivalent to the force $\bar{F}_z$ and a couple that prevents anticlastic bending.

Numerical results are now given in nondimensional form, for two sets of data,

$$\text{(a) } \bar{H} = 0.1,\ \alpha = 0.8 \quad \text{and} \quad \text{(b) } \bar{H} = 0.2,\ \alpha = 1.2.$$

For (a),

$$\bar{a} = 1.20104,\quad \bar{b} = 0.130096,\quad \bar{r}_0 = 1.25,$$
$$\bar{F}_z = -2.1315 \times 10^{-4},\quad \overline{M_z/2W} = 2.66185 \times 10^{-4}.$$

For (b)

$$\bar{a} = 0.73993, \quad \bar{b} = 0.938526, \quad \bar{r}_0 = 0.82768,$$
$$\bar{F}_z = -3.8075 \times 10^{-3}, \quad \overline{M_z/2W} = 3.14635 \times 10^{-3}.$$

It is interesting to compare the above results for M_z with those obtained from linear bending theory. For the linear theory applied to the problem just considereded it follows from the theory given in [13] that $M_z/2W$ is given by

$$M_z/2W = \frac{EH^3}{12\left(1 - v^2\right)} \frac{1}{r_0}, \tag{7.86}$$

where $E = 2\mu(1 + v)$ is Young's modulus, v is Poisson's ratio, and r_0 is radius of curvature of the mid plane. The nondimensional form of equation (7.86) with $v = 0.5$ is

$$\overline{M_z/2W} = \frac{\bar{H}^3}{3\bar{r}_0},$$

which gives

$$\overline{M_z/2W} = 2.667 \times 10^{-4}$$

for (a) and

$$\overline{M_z/2W} = 3.223 \times 10^{-3}$$

for (b). It can be shown that the difference between the values of $\overline{M_z/2W}$ for linear and nonlinear theories increase as $\bar{H}$ and α increase.

7.9 Combined Telescopic and Torsional Shear

In this section we consider a deformation, which is not controllable for an incompressible isotropic hyperelastic solid, although an inverse method of solution of the problem of combined telescopic and torsional shear is applicable. The deformation is assumed to be for an incompressible circular tube, bonded at its outer radius to a fixed rigid sleeve and at its inner radius to a rigid bar that is subjected to prescribed displacement and twist as indicated in Figure 7.7.

This deformation is given by

$$r = R, \quad \theta = \Theta + \alpha(R), \quad z = Z + \mathrm{w}(R), \tag{7.87}$$

in the usual notation for the cylindrical polar coordinates of a material point in the deformed and undeformed states. Equations (7.87) follows from the incompressibility assumption. The boundary conditions are

$$w(A) = \mathrm{w}_i, \quad \alpha(A) = \alpha_i, \tag{7.88}$$

$$\mathrm{w}(B) = \alpha(B) = 0, \tag{7.89}$$

where A and B are the fixed inner and outer radii, respectively, of the tube. The problem is to determine the nonzero components of Cauchy stress so that the axial force and torque required per unit length can be obtained. Instead of equation (7.78) a stress boundary condition

$$\sigma_{rz}(A) = \tau, \; \sigma_{r\theta} = q$$

could be considered and then the problem would be to obtain $\mathrm{w}(R)$ and $\alpha(R)$. It is assumed that the system is sufficiently long that end effects can be neglected.

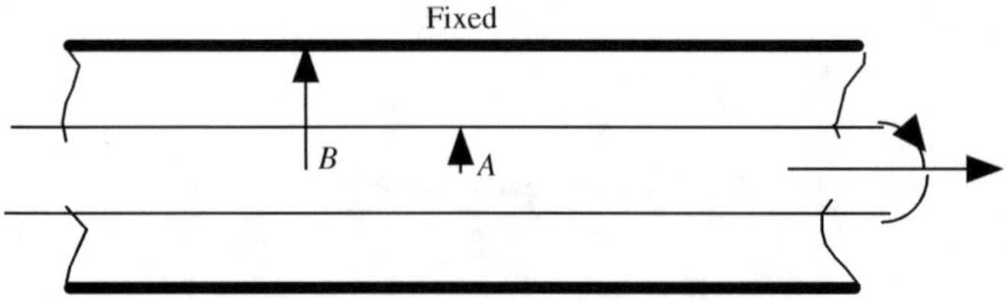

Figure 7.7. Combined telescopic and torsional shear.

Deformation field (7.87) is not controllable since, in general, the functions $r\alpha(R)$ and $\mathrm{w}(R)$ depend on the strain energy function.

If $\alpha(R) = 0$ deformation field (7.87) represents telescopic shear, sometimes known as axial or antiplane shear, and if $\mathrm{w}(R) = 0$, equation (7.87) represents torsional shear sometimes known as azimuthal shear.

The physical components of the deformation gradient $\boldsymbol{F}$, left Cauchy-Green tensor $\boldsymbol{B}$, and $\boldsymbol{B}^{-1}$ are

$$[\boldsymbol{F}] = \begin{bmatrix} 1 & 0 & 0 \\ r\alpha' & 1 & 0 \\ \mathrm{w}' & 0 & 1 \end{bmatrix},$$

$$[\boldsymbol{B}] = \begin{bmatrix} 1 & r\alpha' & \mathrm{w}' \\ r\alpha' & 1 + r^2\alpha'^2 & r\alpha'\mathrm{w}' \\ \mathrm{w}' & r\alpha'\mathrm{w}' & 1 + \mathrm{w}'^2 \end{bmatrix}, \tag{7.90}$$

$$\left[\boldsymbol{B}^{-1}\right] = \begin{bmatrix} 1 + \mathrm{w}'^2 + r^2\alpha'^2 & -r\alpha' & -\mathrm{w}' \\ -r\alpha' & 1 & 0 \\ \mathrm{w}' & 0 & 1 \end{bmatrix},$$

where a prime denotes differentiation with respect to the argument, which can be taken as r since $r = R$.

A solution is outlined for the Mooney-Rivlin solid and the procedure can be adapted to consider other isotropic incompressible hyperelastic solids.

The components of the Cauchy stress tensor obtained from equations (7.5) and (7.90) are

$$\frac{\sigma_r}{\mu} = -\frac{p}{\mu} + \gamma - (1-\gamma)\left(1 + \mathrm{w}'^2 + r^2\alpha'^2\right), \tag{7.91}$$

$$\frac{\sigma_\theta}{\mu} = -\frac{p}{\mu} + \gamma\left(1 + r^2\alpha'^2\right) - (1-\gamma), \tag{7.92}$$

$$\frac{\sigma_z}{\mu} = -\frac{p}{\mu} + \gamma\left(1 + \mathrm{w}'^2\right) - (1-\gamma), \tag{7.93}$$

$$\frac{\sigma_{r\theta}}{\mu} = r\alpha', \tag{7.94}$$

$$\frac{\sigma_{rz}}{\mu} = \mathrm{w}', \tag{7.95}$$

$$\frac{\sigma_{\theta z}}{\mu} = \gamma r\alpha'\mathrm{w}'. \tag{7.96}$$

It is interesting to note that $\sigma_{\theta z} = 0$ when $\gamma = 0$ and for telescopic shear or torsional shear alone. The stresses $\sigma_{r\theta}$ and σ_{rz} do not depend on the telescopic shear and torsional shear, respectively, so that there is partial uncoupling; however, this is only true for the Mooney-Rivlin solid. Also $\sigma_{r\theta}$ and σ_{rz} do not depend on γ.

The equilibrium equations are

$$\frac{\partial\sigma_r}{\partial r} + \frac{\sigma_r - \sigma_\theta}{r} = 0, \tag{7.97}$$

$$\frac{\partial\sigma_{r\theta}}{\partial r} + \frac{2\sigma_{r\theta}}{r} = 0, \tag{7.98}$$

$$\frac{\partial\sigma_{rz}}{\partial r} + \frac{\sigma_{rz}}{r} = 0, \tag{7.99}$$

where r can be replaced by R.

Combining equations (7.94) and (7.98) gives

$$\frac{d^2\alpha}{dr^2} + \frac{3}{r}\frac{d\alpha}{dr} = 0,$$

with solution

$$\alpha = c_1 + c_2 r^{-2},$$

where c_1 and c_2 are constants of integration that are evaluated from the boundary conditions $(7.88)_2$ and $(7.89)_2$ to give

$$\alpha = \alpha_i \frac{\left(r^{-2} - B^{-2}\right)}{\left(A^{-2} - B^{-2}\right)}. \tag{7.100}$$

Similarly combining equations (7.75) and (7.79) gives

$$\frac{d^2 \mathrm{w}}{dr^2} + \frac{1}{r}\frac{d\mathrm{w}}{dr} = 0,$$

with solution

$$\mathrm{w} = -c_3 \ln r + c_4,$$

where c_3 and c_4 are constants of integration that are evaluated from the boundary conditions $(7.84)_1$ and $(7.85)_1$ to give

$$\mathrm{w} = \mathrm{w}_i \ln\frac{r}{B} \Big/ \ln\frac{A}{B}. \tag{7.101}$$

The shearing stresses are then obtained by substituting equations (7.100) and (7.101) in equations (7.94) to (7.96). Since equations (7.100) and (7.101) do not depend on the parameter γ, the deformation is controllable for all Mooney-Rivlin solids.

The normal stress σ_r is obtained by substituting equations (7.100) and (7.101) in equations (7.91) and (7.92) and then substituting equations (7.91) and (7.92) in equation (7.97) and integrating. It is assumed that $\sigma_r = 0$ if $w_i = \alpha_i = 0$, so that the constant of integration is zero and the integral gives

$$\frac{\sigma_r}{\mu} = -\left\{ (1-\gamma)\frac{\mathrm{w}_i^2}{2r^2 (\ln \eta)^2} + \frac{\alpha_i^2 (r/A)^{-4}}{\left(1-\eta^2\right)^2} \right\}. \tag{7.102}$$

Examination of equation (7.102) indicates that a limited super position principle holds for σ_r, that is, σ_r due to w_i applied alone, plus σ_r due to α_i applied alone is equal to σ_r due to w_i and α_i applied together. The form of the Lagrangian multiplier p that appears in equations (7.92) to (7.93) can be obtained from equations (7.91) and (7.102). The normal stresses σ_r, σ_θ, and σ_z are second-order effects for the problem considered. It follows from equation (7.102) that $\sigma_r = 0$ for telescopic shear of the neo-Hookean solid.

EXERCISES

7.1. Solve the thick-walled spherical shell problem for the neo-Hookean solid using the strain energy function in the form

$$w = \breve{w}(\lambda_1, \lambda_2, \lambda_3).$$

7.2. Solve the thick-walled axially symmetric cylindrical tube problem for the neo-Hookean solid using the strain energy function in the form

$$w = \breve{w}(\lambda_1, \lambda_2, \lambda_3).$$

Assume that the z axis of a cylindrical polar system of axes coincides with the axis of the tube and that the corresponding stretch is $\lambda_3 = 1$.

7.3. A thin-walled compressible spherical shell whose strain energy function is given by equation (7.4) has wall thickness T and mean wall radius R in the undeformed reference configuration. Find an approximate expression for the internal pressure at the onset of instability.

7.4. A thick-walled incompressible tube has inner and outer radii A and B, respectively, in the undeformed reference configuration. It spins about its axis with angular velocity ω. Find the relation between the deformed inner radius a and ω, for the neo-Hookean strain energy function. Assume that the length is constrained to remain constant.

REFERENCES

1. Ericksen, J.L. (1954). Deformations Possible in Every Isotropic, Compressible Perfectly Elastic Body. ZAMP 5, pp. 466–489.
2. Ericksen, J.L. (1955). Deformations Possible in Every Isotropic, Incompressible Perfectly Elastic Body. J. Math. Phys. 34, pp. 126–128.
3. Green, A.E., and Zerna, W. (1968).Theoretical Elasticity, 2nd Ed. Oxford University Press.
4. Truesdell, C., Editor. (1965). Continuum Mechanics IV Problems of Non-Linear Elasticity, International Review Series. Gordon and Breach Science Publishers.
5. Ogden, R.W. (1984). Non-Linear Elastic Deformations. Ellis Harwood.
6. Blatz, P.J., and Ko, W.L. (1962). Application of Elastic Theory to the Deformation of Rubbery Materials. Trans. Soc. of Rheology 6, pp. 223–251.
7. Levinson, M., and Burgess, I.W. (1971). A Comparison of Some Simple Constitutive Relations for Slightly compressible Compressible Rubberlike Materials. Int. J. Mech. Sci. 13, pp. 563–572.
8. Poynting, J.H. (1909). On Pressure perpendicular to the Shear Planes in Finite Pure Shears, and on the Lengthening of Loaded Wires When Twisted. Proc. Roy. Soc. London A82, pp. 546–559.
9. Adkins, J.E. (1961). Large Elastic Deformations. Progress in Solid Mechanics, Vol. II. Ed. I.N. Sneddon and R. Hill. North Holland.
10. Beatty, M.F., and Stalnacker, D.O. (1986). The Poisson Function of Finite Elasticity. ASME J. Appl. Mech. 53, pp. 807–813.
11. Rivlin, R.S. (1949). Large Elastic Deformations of Isotropic Materials, Part VI, Further Results in the Theory of Torsion Shear and Flexure. Phil. Trans. Roy. Soc. A.242, pp. 173–195.
12. Wang, C.C., and Truesdell, C. (1973). Introduction to Rational Elasticity. Noordhoff.
13. Timoshenko, S.P., and Woinowsky-Krieger, S. (1959). Theory of Plates and Shells, 2nd Ed. McGraw-Hill.

8 Finite Deformation Thermoelasticity

8.1 Principle of Local State and Thermodynamic Potentials

The theory in this chapter is based on the Principle of Local State given in chapter 4. Kestin [1] has noted that this principle implies that "the local and instantaneous gradients of the thermodynamic properties as well as their local and instantaneous rates of change do not enter into the description of the state and do not modify the equations of state." It follows that the thermodynamic potentials are assumed to be of the same form as for an equilibrium state.

The important potentials in finite deformation thermoelasticity are the Helmholtz free energy and the internal energy, since, as for linear thermoelasticity, the Helmholtz free energy at constant temperature is the isothermal strain energy and the internal energy at constant entropy is the isentropic strain energy. This indicates the relation between hyperelasticity and thermoelasticity.

Application of thermoelasticity to finite deformation of idealized rubber-like materials involves the concept of entropic response and energetic response and this is discussed in this chapter.

8.2 Basic Relations for Finite Deformation Thermoelasticity

The theory in this section is presented in terms of the absolute temperature Θ, the specific entropy s, and the referential variables, $\boldsymbol{S}$, the second Piola-Kirchhoff stress, and $\boldsymbol{C}$, the right Green tensor, or $\boldsymbol{E}$, Green's strain tensor.

There are four thermodynamic potentials that are fundamental equations of state, that is, all the thermodynamic properties can be obtained from a potential by differentiation if the potential is expressed as a function of the appropriate thermodynamic properties. The potentials are related by Legendre transformations and are the specific (per unit mass) internal energy,

$$u = u(\boldsymbol{C}, s), \tag{8.1}$$

the specific Helmholtz free energy,

$$f = f(\boldsymbol{C}, \Theta) = u - \Theta s, \tag{8.2}$$

the specific enthalpy

$$h = h(\boldsymbol{S}, \boldsymbol{s}) = u - \frac{S_{KL}C_{KL}}{2\rho_0}, \tag{8.3}$$

and the specific Gibbs free energy,

$$g = g(\boldsymbol{S},\Theta) = f - \frac{S_{KL}C_{KL}}{2\rho_0}. \tag{8.4}$$

Since $\boldsymbol{C} = \boldsymbol{F}^{\mathrm{T}}\ \boldsymbol{F} = 2\boldsymbol{E} + \boldsymbol{1}$, equations (8.1) and (8.2) can be put in the alternative equivalent forms

$$u = \hat{u}(\boldsymbol{F}, s), \quad f = \hat{f}(\boldsymbol{F},\Theta), \quad u = \widehat{u}(\boldsymbol{E}, s) \quad \text{and } f = \widehat{f}(\boldsymbol{E},\Theta),$$

and equations (8.3) and (8.4) in the equivalent forms

$$h = \hat{h}(\Sigma, \boldsymbol{s}) \text{ and } \mathrm{g} = \mathrm{g}\ (\Sigma,\Theta).$$

The thermodynamic relations based on u and f are important for what follows in this chapter and are now given in terms of the conjugate properties, $\boldsymbol{S}$, the second Piola-Kirchhoff stress, and, $\boldsymbol{C}$, the right Cauchy-Green strain tensor (or $\boldsymbol{E}$ the Lagrangian strain tensor). A Gibbs relation, based on u, that can be deduced from the first and second laws of thermodynamics is

$$\begin{aligned} du &= \frac{S_{KL} dE_{KL}}{\rho_0} + \Theta ds \\ &= \frac{S_{KL} dC_{KL}}{2\rho_0} + \Theta ds. \end{aligned} \tag{8.5}$$

From equation (8.1) or (8.5),

$$du = \frac{\partial u}{dC_{KL}} dC_{KL} + \frac{\partial u}{ds} ds. \tag{8.6}$$

Comparison of equations (8.5) and (8.6) gives

$$S_{KL}(\boldsymbol{C}, s) = 2\rho_0 \frac{\partial u(\boldsymbol{C}, s)}{\partial C_{KL}}, \ \Theta(\boldsymbol{C}, s) = \frac{\partial u(\boldsymbol{C}, s)}{\partial s}. \tag{8.7}$$

A further Gibbs relation, based on f,

$$df = \frac{\tilde{S}_{KL}(\mathbf{C},\Theta) dC_{KL}}{2\rho_0} - s d\Theta, \tag{8.8}$$

follows from equations (8.2) and (8.5). It then follows from equation (8.2) or (8.8) that

$$df = \frac{\partial f}{\partial C_{KL}} dC_{KL} + \frac{\partial f}{\partial \theta} d\Theta. \tag{8.9}$$

Comparison of equations (8.8) and (8.9) gives

$$\tilde{S}_{KL}(\mathbf{C},\Theta) = 2\rho_0 \frac{\partial f(\mathbf{C},\Theta)}{\partial C_{KL}}, \quad s(\mathbf{C},\Theta) = -\frac{\partial f(\mathbf{C},\Theta)}{\partial \Theta}. \tag{8.10}$$

The determination of the Gibbs relations based on h and g is left as an exercise at the end of the chapter.

From equations (8.2) and $(8.10)_2$, the internal energy as a function of $\boldsymbol{C}$ and Θ is given by

$$\tilde{u}(\boldsymbol{C},\Theta) = f(\boldsymbol{C},\Theta) - \Theta\frac{\partial f(\boldsymbol{C},\Theta)}{\partial\Theta}.$$

This is, a useful form, although it is not a fundamental equation of state. It follows from the first law of thermodynamics that at constant deformation $d\tilde{u} = c_c d\Theta \;=\; \Theta ds$, where c_c is the specific heat at constant $\boldsymbol{C}$; consequently,

$$c_c = \frac{\partial\tilde{u}}{\partial\Theta} = \Theta\frac{\partial s(\boldsymbol{C},\,\Theta)}{\partial\Theta}. \tag{8.11}$$

The following relation is obtained from equations (8.2), (8.10), and (8.11)

$$\tilde{S}_{KL}(\boldsymbol{C},\Theta) = 2\rho_0\left(\frac{\partial\tilde{u}}{\partial C_{KL}}\right) - 2\rho_0\Theta\left(\frac{\partial s}{\partial C_{KL}}\right).$$

This shows that the stress consists of two parts, one due to change of internal energy produced by deformation and the other due to the change of entropy. The first part is described as energetic and the second as entropic.

Equations $(8.7)_1$ and $(8.10)_1$ indicate that the strain energy w is given by

$$\rho_0 u(\boldsymbol{C},\,s_0) = w_{isentropic} \text{ and } \rho_0 f(\boldsymbol{C},\,\Theta) = w_{isothermal},$$

for isentropic deformation and isothermal deformation, respectively, with reference temperature and entropy Θ_0 and s_0.

An explicit expression for $f(\boldsymbol{C},\,\Theta)$ is obtained as follows. Combining gives equations $(8.10)_2$ and equation $(8.11)_2$ gives

$$c_c(\boldsymbol{C},\Theta) = -\Theta\frac{\partial^2 f(\boldsymbol{C},\Theta)}{\partial\Theta^2},$$

and integrating once from Θ_0 to Θ gives

$$\left(\frac{\partial f}{\partial \Theta}\right) - \left(\frac{\partial f}{\partial \Theta}\right)_{\Theta=\Theta_0} = -\int_{\Theta_0}^{\Theta} \frac{c_c(\boldsymbol{C}, \Theta')}{\Theta'} d\Theta'. \tag{8.12}$$

Equations (8.2) and $(8.10)_2$ give

$$f = \tilde{u} + \Theta \frac{\partial f}{\partial \Theta}, \tag{8.13}$$

and it can be deduced from this that

$$\left(\frac{\partial f}{\partial \Theta}\right)_{\Theta=\Theta_0} = \frac{1}{\Theta_0}\left\{f\left(\boldsymbol{C}, \boldsymbol{\Theta_0}\right) - \tilde{u}\left(\boldsymbol{C}, \boldsymbol{\Theta_0}\right)\right\},$$

and substitution in equation (8.12) gives

$$\frac{\partial f}{\partial \Theta} = \frac{1}{\Theta_0}\left\{f\left(\boldsymbol{C}, \Theta_0\right) - \tilde{u}\left(\boldsymbol{C}, \Theta_0\right)\right\} - \int_{\Theta_0}^{\Theta} \frac{c_c(\boldsymbol{C}, \Theta')}{\Theta'} d\Theta'. \tag{8.14}$$

Integration of equation $(8.11)_1$ from Θ_0 to Θ gives

$$\tilde{u}(\boldsymbol{C}, \Theta) = \int_{\Theta_0}^{\Theta} c_c(\boldsymbol{C}, \Theta') d\Theta' + \tilde{u}(\boldsymbol{C}, \boldsymbol{\Theta_0}). \tag{8.15}$$

The relation

$$\begin{aligned} f(\boldsymbol{C}, \Theta) = {} & \frac{\Theta}{\Theta_0} f(\boldsymbol{C}, \Theta_0) - \left(\frac{\Theta}{\Theta_0} - 1\right)\tilde{u}(\boldsymbol{C}, \Theta_0) - \\ & \int_{\Theta_0}^{\Theta} \left(\frac{\Theta}{\Theta'} - 1\right) c_c(\boldsymbol{C}, \Theta') d\Theta' \end{aligned} \tag{8.16}$$

is then obtained by substituting equations (8.14) and (8.15) in equation (8.13).

The Clausius-Duhem or entropy inequality (4.38) can be put in the form

$$\rho_0(\Theta\dot{s} - \dot{u}) + \tilde{S}_{KL}\dot{C}_{KL}/2 - \frac{Q_K}{\Theta}\frac{\partial\Theta}{\partial X_K} \geqslant 0, \tag{8.17}$$

where $\boldsymbol{Q}$ is the referential heat flux that is related to the spatial heat flux, $\boldsymbol{q}$, by

$$\boldsymbol{q} = J^{-1}\boldsymbol{F}\boldsymbol{Q}.$$

The alternative form,

$$-\rho_0\left(s\dot{\theta} + \dot{f}\right) + \tilde{S}_{KL}\dot{C}_{KL}/2 - \frac{Q_k}{\Theta}\frac{\partial\Theta}{\partial X_K} \geqslant 0, \tag{8.18}$$

is obtained from equation (8.2). Substituting

$$\dot{u} = \frac{\partial u}{\partial C_{KL}}\dot{C}_{KL} + \frac{\partial u}{\partial s}\dot{s},$$

which follows from equation (8.6), in equation (8.17) gives the entropy inequality for a thermoelastic solid,

$$\left(\frac{1}{2}S_{KL} - \rho_0\frac{\partial u}{\partial C_{KL}}\right)\dot{C}_{KL} + \rho_0\left(\Theta - \frac{\partial u}{\partial s}\right)\dot{s} - \frac{Q_K}{\theta}\frac{\partial\Theta}{\partial X_K} \geqslant 0. \tag{8.19}$$

Substituting

$$\dot{f} = \frac{\partial f}{\partial C_{KL}}\dot{C}_{KL} + \frac{\partial f}{\partial \theta}\dot{\Theta},$$

that follows from equation (8.9), into equation (8.18) gives an alternative form

$$\left(\frac{1}{2}\tilde{S}_{KL} - \rho_0 \frac{\partial f}{\partial C_{KL}}\right)\dot{C}_{KL} - \rho_0\left(s + \frac{\partial f}{\partial \Theta}\right)\dot{\Theta} - \frac{Q_K}{\Theta}\frac{\partial \Theta}{\partial X_K} \geqslant 0. \tag{8.20}$$

Inequality (8.20) must hold for all values of $\dot{\boldsymbol{C}}$ and $\dot{\Theta}$, and this gives an alternative derivation of equations (8.10) since the factors of $\dot{\boldsymbol{C}}$ and $\dot{\Theta}$ in equation (8.20) must be zero. The same reasoning applied to equations (8.19) gives an alternative derivation of equations (8.7).

It follows from equations (8.7), and (8.19) or (8.10) and (8.20) that

$$\dot{\gamma} = -\frac{Q_K}{\rho_0\Theta^2}\frac{\partial \Theta}{\partial X_K} \geqslant 0, \text{ or } -Q_K\frac{\partial \Theta}{\partial X_K} \geqslant 0, \tag{8.21}$$

where $\dot{\gamma}$ is the rate of entropy production per unit mass. As an exercise the reader should show that the spatial forms of equations (8.21) are

$$\dot{\gamma} = -\frac{q_k}{\rho\Theta^2}\frac{\partial \Theta}{\partial x_k} \geqslant 0 \text{ or } -q_K\frac{\partial \Theta}{\partial x_k} \geqslant 0.$$

Inequality $(8.21)_2$ is consistent with the Clausius statement of the second law of thermodynamics that states that "heat can cannot pass spontaneously from a body of lower temperature to a body of higher temperature."

Entropy is produced in a thermoelastic solid by temperature gradients resulting in heat conduction, a thermodynamically irreversible process, with entropy production as indicated by equation (8.21), when shock waves occur there is an entropy jump across a shock. There is no entropy production in an ideal thermoelastic solid due to hysteresis effects.

Usually in elastostatic problems thermal effects are neglected. This leads to consideration of the piezotropic model for which mechanical and thermal effects are uncoupled. For this model,

$$u(\boldsymbol{C}, s) = u_1(\boldsymbol{C}) + u_2(s),$$

and it may be deduced that the volume coefficient of thermal expansion is zero for a piezotropic solid. The zero coefficient of thermal expansion indicates that, in an exact sense, the piezotropic model is not physical; however, its use is implied when thermal effects are neglected and the hyperelastic model is assumed for static problems. Then, since the mechanical effects are uncoupled it may be shown from conservation of mechanical energy that the specific stress power is equal to the material time derivative of the strain energy.

8.3 Ideal Rubber-Like Materials

Material in this section is influenced by important papers by Chadwick [2] and Chadwick and Creasy [3] that should be referred to for a more detailed treatment of the topic.

Most solid materials that can undergo finite elastic strain are polymers or rubber-like materials, henceforth described as rubbers. These are composed of entanglements of high–molecular weight chains. This is in contrast to hard solids such as metals that are essentially crystalline. The stresses in an ideal rubber result from changes in entropy, only, and these solids are said to be strictly entropic. Most nonideal rubbers are not strictly entropic but are partly energetic and are said to exhibit modified entropic elasticity [3]. In particular this is true for compressible elastic solids. A suitable model for a compressible isotropic rubber-like elastic solid has entropic behavior for pure shear and energetic behavior for dilatation so that the internal energy can be expressed in the form

$$\tilde{u} = \tilde{u}_1(J) + \tilde{u}_2(\Theta), \tag{8.22}$$

where $J = \det[\boldsymbol{F}] = \sqrt{\det \boldsymbol{C}}$. Henceforth, equation (8.22) is assumed in this section.

Equations $(8.11)_1$ and (8.22) imply that the specific heat at constant deformation is a function of temperature only or constant.

The bulk modulus K for infinitesimal deformation of most rubbers is up to three orders of magnitude greater than the corresponding shear modulus μ; however, the volume coefficient of thermal expansion is about three times a typical value for a metal. This has motivated a model that is mechanically incompressible but has a nonzero coefficient of volume thermal expansion [4]. We avoid the use of this model since it is thermodynamically inconsistent. It may be reasonable in a thermoelastic problem involving a rubber-like material to neglect the dilatation, due to mechanical compressibility, compared with that due to volume thermal expansion and this is essentially the same as assuming the mechanically incompressible model with a nonzero coefficient of volume thermal expansion. However, it is possible that the dilatation due to the isotropic part of the stress tensor could be of the same order of magnitude as that due to thermal expansion depending on the particular problem.

In this section we consider both incompressible and compressible isotropic models for rubber-like solids and express the deformation in terms of the principal stretches λ_i, $i \in \{1, 2, 3\}$. In the present discussion an incompressible solid is defined as a solid capable of isochoric deformation only so that the volume coefficient of thermal expansion is zero. This model is thermodynamically consistent but not strictly physical. The temperature in the natural reference configuration is denoted by Θ_0, and it is assumed that $|\Theta - \Theta_0|/\Theta_0 < < 1$ and $|J - 1 < < 1|$, for the problems considered.

A physically realistic form for the isothermal strain energy function per unit volume of the natural reference configuration, for an isotropic compressible rubber rubber-like solid at temperature Θ_0 is [2],

$$w(\lambda_i) = \mu\phi(\lambda_i) + Kg(J), \tag{8.23}$$

where μ and K are the isothermal shear and bulk moduli, respectively, for infinitesimal deformation from the natural reference configuration, ϕ is a nondimensional symmetric function of the principal stretches, λ_i, $i \in \{1, 2, 3\}$, which vanishes when $\lambda_1 = \lambda_2 = \lambda_3$, and g is a nondimensional function of $J = \lambda_1\lambda_2\lambda_3$ that satisfies the conditions $g(1) = 0$, $g'(1) = 0$ *and* $g'' = 1$, where a prime denotes differentiation with respect to J. It may be shown that the compressible models given by equations (7.3) and (7.4) can be put in the form (8.22). For most rubbers the ratio

μ/K is of order of magnitude 10^{-3}, and it follows that for finite distortion the dilatation may be considered infinitesimal unless the isotropic part of the Cauchy stress is very large compared with μ, Ogden [5] has proposed an empirical relation

$$p = \frac{K}{9}\left(J^{-1} - J^{-10}\right), \tag{8.24}$$

where $p = -\sigma_{kk}/3$, and, since, $p = -K\partial g(J)/\partial K$ for hydrostatic compression,

$$g = \frac{1}{9}\left\{\frac{J^{-9}}{9} + \ln J - \frac{1}{9}\right\}. \tag{8.25}$$

For $|J - 1|$ of the same order of magnitude as for classical elasticity, equations (8.24) and (8.25) could be replaced by

$$p = -K(J - 1), \tag{8.26}$$

and

$$g = \frac{1}{2}(J - 1)^2, \tag{8.27}$$

respectively, with negligible difference from equations (8.24) and (8.25) for $0.98 < J < 1.02$. Consequently, equations (8.26) and (8.27) are assumed in what follows.

Since, for the problems considered, it is assumed that $|\Theta - \Theta_0|/\Theta_0 << 1$ the approximation $c_c =$ constant is adopted, and from equations $(8.11)_1$ and (8.22),

$$\tilde{u}_2 = c_c(\Theta - \Theta_0),$$

where, as before, c_c is the specific heat at constant deformation. Also equation (8.16) is then approximated in terms of the principal stretches, for a rubber-like elastic solid, by

$$f(\lambda_i, \Theta) = \frac{\Theta}{\Theta_0} f(\lambda_i, \Theta_0) - \left(\frac{\Theta}{\Theta_0} - 1\right)\tilde{u}_1(J) - c_c\Theta \ln\frac{\Theta}{\Theta_0} + c_c(\Theta - \Theta_0). \tag{8.28}$$

It has been shown, from the statistical theory of rubber elasticity [6], that an ideal rubber is incompressible, is strictly entropic, that is, the internal energy can be expressed as a function $u = u(\Theta)$ of temperature, only, as for a perfect gas and has the Helmholtz free energy, at temperature Θ_0, given, by

$$w/\rho_0 = f(\lambda_i, \Theta_0) = \frac{\mu}{2\rho_0}\left(\lambda_1^2 + \lambda_2^2 + \lambda_3^2 - 3\right), \quad \lambda_1\lambda_2\lambda_3 = 1. \tag{8.29}$$

In equation (8.29), w is the neo-Hookean strain energy function; however, the following theory can be applied to other admissible incompressible strain energy functions. The neo-Hookean model gives results for simple tension of most rubber that are in good agreement with experiments for an axial stretch up to about 1.8. Numerical results for simple tension of the neo-Hookean model and a compressible generalization are given in this section for an axial stretch up to 1.6.

In order to consider the consequences of strictly entropic elasticity we consider simple tension with axial stretch λ and axial nominal stress $\Sigma = \sigma/\lambda$, where σ is the corresponding axial Cauchy stress.

For simple tension, the axial nominal stress is identical to the axial Biot stress S^B. For isothermal simple tension equation (8.29) becomes

$$w = \rho_0 f(\lambda, \Theta_0) = \frac{\mu}{2}\left(\lambda^2 + 2/\lambda - 3\right), \tag{8.30}$$

and

$$\Sigma = \frac{dw}{d\lambda} = \mu\left(\lambda - \lambda^{-2}\right), \quad \sigma = \lambda\frac{dw}{d\lambda} = \mu\left(\lambda^2 - \lambda^{-1}\right). \tag{8.31}$$

The simple tension stress (8.31) is in good agreement with experimental results for stretch up to about 1.8.

Since $u_2(\Theta_0) = 0$ and $J = 1$, equation (8.28) becomes

$$f(\lambda, \Theta) = \frac{\Theta}{\Theta_0} f(\lambda, \Theta_0) - c\Theta \ln \frac{\Theta}{\Theta_0} + c_c(\Theta - \Theta_0). \tag{8.32}$$

The nominal stress obtained from equations $(8.31)_1$ and (8.32) is then given by

$$\Sigma(\lambda,\Theta) = \rho_0 \frac{\Theta}{\Theta_0} \frac{df(\lambda,\Theta_0)}{d\lambda} = \frac{\Theta}{\Theta_0} \frac{dw}{d\lambda}, \tag{8.33}$$

which shows that Σ is proportional to the absolute temperature for a given value of λ. It follows from equation $(8.33)_2$ that if the temperature of the reference state is increased and deformation constrained, the stress will remain zero. This means that, for a strictly entropic model for an elastic solid the coefficient of volume thermal expansion is zero.

The entropy is given by

$$s = -\frac{\partial f(\lambda,\Theta)}{\partial \Theta} = -\left\{ \frac{f(\lambda,\Theta_0)}{\Theta_0} - c_c \ln \frac{\Theta}{\Theta_0} \right\}, \tag{8.34}$$

which shows that s is of the form $s = s_1(\lambda) + s_2(\Theta)$. It follows from equations (8.33) and (8.34) that, for isothermal simple tension,

$$\Sigma = -\rho_0 \Theta \frac{\partial s}{\partial \lambda},$$

and this indicates that the stress is obtained from the change of entropy with stretch for isothermal deformation.

8.4 Isentropic Simple Tension of Ideal Rubber

We now consider isentropic simple tension from the reference state. It may be deduced from equation (8.34) that

$$\Theta = \Theta_0 \exp\left(\frac{f(\lambda,\Theta_0)}{c_c\Theta_0}\right), \tag{8.35}$$

since $s = 0$ and $\lambda = 1$ in the reference state, and a temperature rise

$$\Theta - \Theta_0 = \Theta_0\left[\exp\left(\frac{w(\lambda)}{\rho_0 c\Theta_0}\right) - 1\right]$$

results from the deformation. The relation between λ and $\Theta - \Theta_0$ is shown graphically in Figure 8.1 for $w(\lambda)$ given by equation (8.30), and

$$\rho = 906.5\ kg.m^{-3}, \quad \mu = 420kPa \text{ and}$$
$$c_c = 1662\ J.kg^{-1}\mathrm{K}^{-1}, \quad \Theta_0 = 293.15\mathrm{K},$$

where K denotes degrees Kelvin. These properties are given in [2].

The rise in temperature of an elastic band due to finite axial strain was first observed by Gough in 1805 and studied experimentally by Joule in 1859. This temperature rise is in contrast to the decrease in temperature

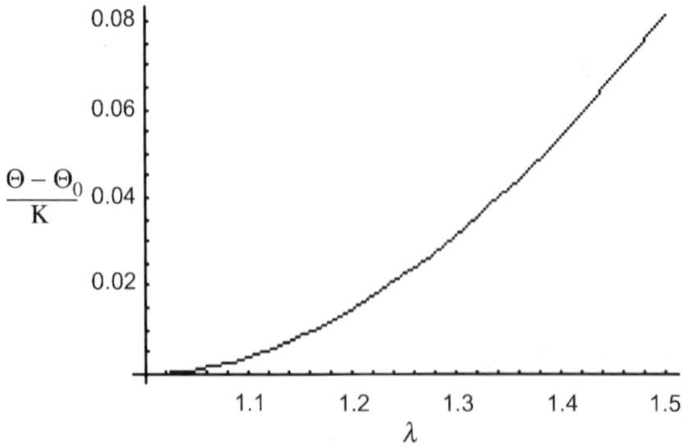

Figure 8.1. Temperature–stretch relation for isentropic simple tension of ideal (incompressible) rubber.

that occurs when a classically thermoelastic bar is subjected to isentropic simple tension where the stress is due to change of internal energy.

The uniaxial nominal (or Biot) stress for isentropic simple tension of an ideal rubber

$$\Sigma = \exp\left(\frac{w}{\rho_0 c_c \Theta_0}\right)\left[\mu\left(\lambda - \lambda^{-2}\right)\right],$$

follows from equations (8.33) and (8.34).

It should be noted that the strictly entropic ideal rubber model is a considerable idealization and the temperature rise indicated in Figure 8.1 is unrealistically high for a given value of λ. However, the above use of this model illustrates some important aspects of entropic thermoelasticity.

8.5 Isentropic Simple Tension of Compressible Rubber

Isentropic simple tension of a compressible model for an elastic rubber-like material with a positive volume coefficient of thermal expansion exhibits a decrease in temperature from $\lambda = 1$ to a minimum, known as the thermoelastic inversion point, at approximately $\lambda = 1.06$. The temperature then increases monotonically, as indicated in Figure 8.2, with zero temperature change at approximately $\lambda = 1.125$. A mechanically incompressible thermoelastic model, with a finite positive volume coefficient of thermal expansion has been used by Holzapfel [4] to analyze this inversion relation. In the present treatment this model is not considered.

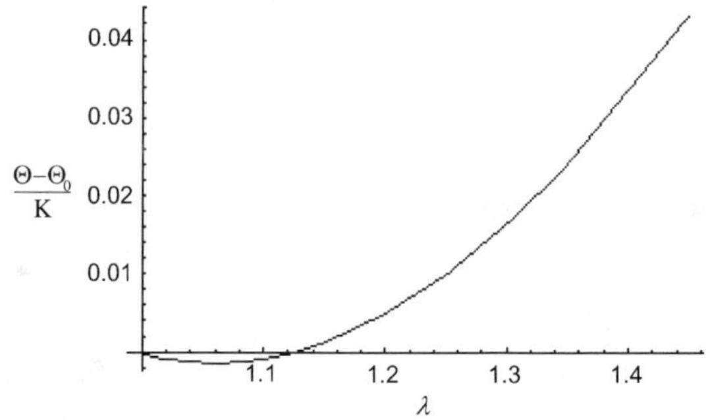

Figure 8.2. Isentropic simple tension of compressible rubber.

In order to investigate the thermoelastic inversion effect in isentropic simple tension of a rubber, for which equation (8.22) is valid, we consider a special case of equation (8.23) that is a compressible generalization of the neo-Hookean strain energy function,

$$w = \rho_0 f(\lambda_i, \Theta_0) = \frac{\mu}{2}\left(\lambda_1^2 + \lambda_2^2 + \lambda_3^2 - 3J^{2/3}\right) + \frac{1}{2}K(J-1)^2. \tag{8.36}$$

Before considering isentropic simple tension we first consider a general homogeneous deformation, then, with the use of equation (8.22), equation (8.28) can be put in the form

$$f(\lambda_i, \Theta) = \frac{\Theta}{\Theta_0}\frac{w}{\rho_0} - \left(\frac{\Theta}{\Theta_0} - 1\right)u_1(J) - c_c\Theta \ln\frac{\Theta}{\Theta_0} + c_c(\Theta - \Theta_0), \tag{8.37}$$

where w is given by equation (8.36). The entropy is then given by

$$s = -\frac{\partial f}{\partial \Theta} = -\left\{\frac{w}{\rho_0\Theta_0} - \frac{u_1(J)}{\Theta_0} - c_c \ln\frac{\Theta}{\Theta_0}\right\}. \tag{8.38}$$

and the principal Biot stresses by

$$S_i^B = \rho_0\frac{\partial f}{\partial \lambda_i} = \frac{\Theta}{\Theta_0}\frac{\partial w}{\partial \lambda_i} - \rho_0\left(\frac{\Theta}{\Theta_0} - 1\right)\frac{d\tilde{u}(J)}{dJ}\frac{\partial J}{\partial \lambda_i}.\ i \in \{1, 2, 3\}. \tag{8.39}$$

In equations (8.37) to (8.39) the term $u_1(J)$ is approximated by

$$u_1 = \frac{\alpha K \Theta_0}{\rho_0}(J-1), \tag{8.40}$$

where α is the volume coefficient of thermal expansion, assumed to be constant for the range of temperature involved. Equation (8.40) is obtained from equation (5.37) and the assumption that the dilatational strain is the same order of magnitude as in classical elasticity. Instead of the factor $(J-1)$, in equation (8.40), Chadwick [2] has proposed $0.4\left(J^{2.5} - 1\right)$, but the difference is negligible for $0.98 < J < 1.02$.

We now consider a particular deformation, simple tension, and outline a procedure that requires numerical implementation for a particular set of data. For simple tension $\lambda = \lambda_1$, $\lambda_2 = \lambda_3 = \lambda_r$, S_1^B is the uniaxial Biot (or nominal stress) and $S_2^B = S_3^B = 0$.

A relation

$$\phi(\lambda_r, \lambda, \Theta) = 0 \tag{8.41}$$

is obtained from $S_2^B = 0$, where, for isothermal simple tension Θ is replaced by Θ_0. For isentropic simple tension from the natural reference state, with $J = 1$ and $s = 0$,

$$\left\{\frac{w}{\rho_0\Theta_0} - \frac{u_1(J)}{\Theta_0} - c_c \ln\frac{\Theta}{\Theta_0}\right\} = 0, \tag{8.42}$$

follows from equation (8.38). Substitution of

$$\Theta = \Theta_0 \exp\left\{\left(\frac{w}{\rho_0 c_c \Theta_0} - \frac{u_1(J)}{\Theta_0}\right)\Big/ c_c\right\} \tag{8.43}$$

obtained from equation (8.42) in equation (8.41) gives a relation between λ_r and λ, for isentropic simple tension, that, for a given value of λ, can be solved for λ_r. The temperature is then determined for the given value of λ by substituting λ_r in equation (8.43). The axial stress is determined, for the given value of λ, by substituting equation (8.43) in S_1^B obtained from equation (8.39), then substituting the already determined value of λ_r. Implementation of this procedure was done using Mathematica. Without the use of Mathematica or a similar package implementation would involve considerable computational effort.

Numerical results for isentropic simple tension that are now presented are for the following data given in [2],

$$\rho_0 = 906.5\,\text{kg.m}^{-3}, \quad \mu = 420\,\text{kPa}, \quad c_c = 1662\,\text{J.kg}^{-1}\text{K}^{-1},$$
$$K = 1.95 \times 10^6\, kPa, \quad \alpha = 6.36 \times 10^{-4}\,\text{K}^{-1}, \quad \Theta_0 = 293.15.$$

The temperature change as a function of λ for isentropic simple tension is shown graphically in Figure 8.2.
The inversion effect, namely, the initial decrease in temperature to a minimum followed by a monotonic increase in temperature is evident in Figure 8.2. This effect is well known and has been extensively studied.

There is negligible difference between results from the incompressible stress-stretch relation (8.31) and those for the isothermal and isentropic relations for the corresponding compressible strain energy function (8.36) with $K/\mu = 4642.9$ obtained from the above data. For example, for $\lambda = 1.6$,

$$\Sigma/\mu = 1.209375, \quad 1.20686, \quad 1.21226,$$

for equation (8.31), isothermal compressible and isentropic compressible, respectively. The corresponding values for λ_r are

$$\lambda_r = 0.790569, \quad 0.790601, \quad 0.790622.$$

The compressible model based on equation (8.22) is a special case of a modified entropic elasticity with the dilatational part of the deformation is energetic and the isochoric part is entropic. A more general modified entropic model has the isochoric part of the deformation partly entropic and partly energetic and for this model (8.22) is replaced by

$$\tilde{u}(\boldsymbol{F},\Theta) = u_1(\boldsymbol{F}) + u_2(\Theta).$$

There is some experimental evidence that this model is applicable to certain rubber-like materials. A detailed discussion of modified entropic elasticity is given in [3].

EXERCISES

8.1. Determine the Gibbs relations for the enthalpy, $h = h(\boldsymbol{S}, s)$, and the Gibbs free energy, $g = g(\boldsymbol{S},\Theta)$.

8.2. Prove that

$$\Theta = \frac{\partial h}{\partial s}, \quad C_{KL} = -2\rho_o \frac{\partial h}{\partial S_{KL}}, \quad s = -\frac{\partial g}{\partial \Theta}, \quad C_{KL} = -2\rho_0 \frac{\partial g}{\partial S_{KL}}.$$

8.3. Prove that $f - u + h - g = 0$.

REFERENCES

1. Kestin, J. (1968). A Course in Thermodynamics, 2nd Ed. Blaisdell Publishing Company.
2. Chadwick, P. (1974). Thermo-Mechanics of Rubber-like Materials. Phil. Trans. R. Soc. A. 276, pp. 371–403.
3. Chadwick, P., and Creasy, C.F.M. (1984). Modified Entropic Elasticity of Rubberlike Materials. J. Mech. Phys. Solids. 32, No. 5, pp. 337–357.
4. Holzapfel, G.A. (2000). Nonlinear Solid Mechanics. Wiley.
5. Ogden, R.W. (1997). Non-Linear Elastic Deformations. Dover.
6. Treloar, L.R.G. (1956). The Physics of Rubber Elasticity, 2nd Ed. Oxford.

9 Dissipative Media

Ideal elastic solids and inviscid fluids do not exhibit dissipative effects. Consequently, the only irreversible processes possible in these media are heat conduction and shock wave propagation. Deformation and flow of dissipative media involve dissipation, due to hysteresis effects, of mechanical energy as heat and this is an additional contribution to irreversibility. Entropy production is a measure of irreversibility, and for a dissipative medium, it has contributions from dissipation of mechanical energy, heat conduction, and shock wave propagation. We will not be concerned with shock wave propagation in this chapter. This topic is covered in the classic text by Whitham [1].

The dissipative media discussed in this chapter are Newtonian viscous fluids, Stokesian fluids, and linear viscoelastic solids. Other dissipative media are discussed in texts on rheology, for example, Tanner [2].

9.1 Newtonian Viscous Fluids

The constitutive equations of a Newtonian viscous fluid have been obtained in chapter 5. In the present chapter the dissipation of mechanical energy and the related entropy production are considered. The constitutive equation (5.10), obtained in chapter 5, for a compressible Newtonian viscous fluid can be put in the form

$$\begin{aligned} \boldsymbol{\sigma} + p\boldsymbol{I} &= 2\eta\boldsymbol{D} + \left\{\left(\xi - \frac{2}{3}\eta\right)\mathrm{tr}\boldsymbol{D}\right\}I, \\ \sigma_{ij} + p\delta_{ij} &= 2\eta d_{ij} + \left\{\left(\xi - \frac{2}{3}\eta\right)d_{kk}\right\}\delta_{ij}, \end{aligned} \tag{9.1}$$

where p is the thermodynamic pressure and η and ξ are the shear and bulk viscosity coefficients, respectively, which are temperature dependent in real fluids. The shear viscosity η of a gas increases with temperature increase and the viscosity of a liquid decreases with temperature increase. Temperature dependences of η and ξ are neglected in what follows, and this restricts the theory presented to relatively small temperature changes. Constitutive equation (9.1) can be decomposed into dilatational and deviatoric parts

$$\frac{\sigma_{kk}}{3} + p = \xi d_{kk} \tag{9.2}$$

and

$$s_{ij} = 2\eta d'_{ij}. \tag{9.3}$$

Equation (9.2) is obtained by contracting equation (9.1), and equation (9.3) is obtained by substituting

$$\begin{aligned} \sigma_{ij} &= s_{ij} + \frac{\sigma_{kk}}{3}\delta_{ij}, \\ d_{ij} &= d'_{ij} + \frac{d_{kk}}{3}\delta_{ij}, \end{aligned} \tag{9.4}$$

into equation (9.1) and using equation (9.2).

It follows from equations (9.2) and (9.3) that the stress power per unit mass consists of a dilatational part and a deviatoric part

$$\frac{\left(\xi d_{kk}^2 - p d_{kk}\right)}{\rho} \quad \text{and} \quad \frac{2\mu d'_{ij} d'_{ij}}{\rho}, \tag{9.5}$$

respectively. The stress power per unit mass

$$\frac{\sigma_{ij} d_{ij}}{\rho} = \frac{1}{\rho}\left(-p d_{kk} + 2\eta d_{ij} d_{ij} + \left(\xi - \frac{2}{3}\eta\right) d_{kk}^2\right)$$

can also be obtained from equation (9.1) and substitution of equation (9.4) gives, after some manipulation,

$$\frac{\sigma_{ij}d_{ij}}{\rho} = \frac{1}{\rho}\left(-pd_{kk} + 2\eta d'_{ij}d'_{ij} + \xi d^2_{kk}\right). \tag{9.6}$$

In equation (9.6) the stress power has been decomposed into three parts, a nondissipative dilatational part pd_{kk}/ρ, and dissipative parts, $2\eta d'_{ij}d'_{ij}/\rho$ due to shear deformation and $\xi d^2_{kk}/\rho$ due to dilatation. This is consistent with equation (9.5). The dissipative parts are rates of dissipation per unit mass of mechanical energy as heat. It follows that the rates of entropy production per unit mass due to shear deformation and dilatation are

$$\dot{\gamma}_s = \frac{2\eta d'_{ij}d'_{ij}}{\rho\Theta} \text{ and } \dot{\gamma}_\text{d} = \frac{\xi d^2_{kk}}{\rho\Theta},$$

respectively. In addition there will be the rate of entropy production due to heat conduction,

$$\dot{\gamma}_c = -\frac{q_i}{\rho\Theta^2}\frac{\partial\Theta}{\partial x_i}.$$

If the Stokes' hypothesis, that is, $\xi = 0$, is assumed, the thermodynamic pressure is $p = -\sigma_{kk}/3$, and there is no viscous dissipation of mechanical energy as heat in dilatation. According to Schlichting [3], it appears that the bulk viscosity vanishes identically in gases of low density.

If an incompressible fluid model is considered, equation (9.1) is replaced by

$$\sigma_{ij} + p\delta_{ij} = 2\eta d_{ij}, \; d_{kk} = 0, \tag{9.7}$$

where p is a Lagrangian multiplier determined from the stress boundary conditions. The trace of equation (9.7) shows that $p = -\text{tr}\boldsymbol{\sigma}/3$. Consequently, equation (9.7) can be put in the form

$$s_{ij} = 2\eta d_{ij}, \quad d_{kk} = 0. \tag{9.8}$$

It is evident that only the shear viscosity coefficient is required.

Equation (9.8) indicates that a dissipation potential, $\phi = (\eta d_{ij} d_{ij})/\rho$, $d_{kk} = 0$, exists, that is , half the rate of dissipation of mechanical energy or stress power $(2\eta d_{ij} d_{ij})/\rho$ per unit mass, so that

$$s_{ij} = \rho \frac{\partial \phi}{\partial d_{ij}}. \tag{9.9}$$

In equation (9.9), as usual, when an isotropic scalar function of a second-order symmetric tensor is differentiated with respect to a component of the tensor, the nine components of the tensor are taken as independent.

9.2 A Non-Newtonian Viscous Fluid

We now consider a model for a nonlinear viscous fluid. This model is an incompressible variant of one known as the Reiner-Rivlin fluid or Stokesian fluid [4] and is given by

$$\boldsymbol{\sigma} = -p\boldsymbol{I} + \alpha_1 \boldsymbol{D} + \alpha_2 \boldsymbol{D}^2, \mathrm{tr}\boldsymbol{D} = 0, \tag{9.10}$$

$$\sigma_{ij} = -p\delta_{ij} + \alpha_1 d_{ij} + \alpha_2 d_{ik} d_{kj}, d_{kk} = 0,$$

where p is determined from the stress boundary conditions. The parameters α_1 and α_2 are functions of the invariants of $\boldsymbol{D}$ and possibly temperature and $\alpha_1/2$ can be regarded as the viscosity and α_2 as the cross viscosity. Since incompressibility is assumed, $\boldsymbol{D}$ is a deviatoric, that is, $\boldsymbol{D} = \boldsymbol{D}'$. It follows that the nonzero invariants of $\boldsymbol{D}$ are

$$\boldsymbol{I}_1 = 0, \quad I_2 = -\frac{1}{2}\mathrm{tr}\boldsymbol{D}'^2, \quad I_3 = \mathrm{Det}[\boldsymbol{D}'] = \frac{1}{3}\mathrm{tr}\boldsymbol{D}'^3.$$

Equation (9.10) indicates that $\boldsymbol{\sigma} + p\boldsymbol{I}$ is due only to viscosity effects. Consequently, the stress power per unit volume

$$P = \alpha_1 \mathrm{tr}\boldsymbol{D}^2 + \alpha_2 \mathrm{tr}\boldsymbol{D}^3$$

is equal to the rate dissipation of mechanical energy. It then follows that

$$\alpha_1 \mathrm{tr}\boldsymbol{D}^2 + \alpha_2 \mathrm{tr}\boldsymbol{D}^3 \geqslant 0$$

is a necessary condition for equation (9.10) to be admissible.

The incompressible Newtonian fluid is a special case of equation (9.10). Equation (9.10) predicts second-order effects that are not predicted by equation (9.8), for example, the normal stresses that are predicted for simple shear flow. However, certain other phenomena, observed for nonlinear dissipative fluids, are not predicted by equation (9.10), and a viscoelastic fluid model is then required.

9.3 Linear Viscoelastic Medium

A viscoelastic medium, usually a solid, exhibits creep under constant stress and stress relaxation under constant deformation. Similar to a viscous fluid the stress is rate dependent but also elastic energy can be stored and dissipated. The linear theory is based on infinitesimal strain, $\boldsymbol{e}$, as in linear elasticity and $\dot{\boldsymbol{e}} \equiv d\boldsymbol{e}/dt \approx \partial \boldsymbol{e}/\partial t$ is described as the strain rate, rather than the rate of deformation when finite deformation is considered. In what follows isotropy is assumed. A viscoelastic solid behaves elastically when the rate of deformation is vanishingly small and the corresponding elastic moduli are known as the relaxation moduli. The relaxation modulus for a viscoelastic fluid is zero. Elastic behavior is approached as the rate of deformation approaches infinity and the corresponding elastic moduli are known as the impact moduli. In this section an outline of linear viscoelasticity is given. Detailed treatments, of the topic, are given in various specialized texts by Christensen [5], Flügge [6], and others.

One-dimensional viscoelastic constitutive relations can be obtained from the study of the response of spring-dashpot models. These models consist of discrete representations of uniform bars subjected to simple tension and compression. A linear spring that schematically represents a uniform elastic bar is shown in Figure 9.1.

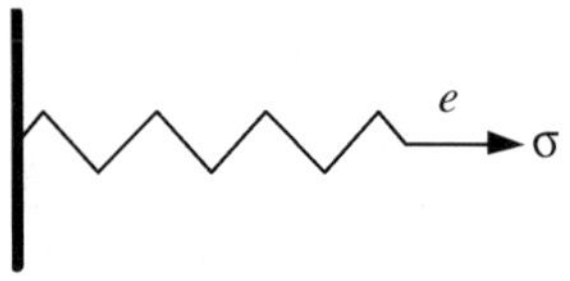

Figure 9.1. Elastic element.

The uniaxial stress and strain are denoted by σ and e, respectively, and are related by Hooke's law,

$$\sigma = Ee, \tag{9.11}$$

where E is constant. A linear viscous element that is modeled schematically by a dashpot is shown in Figure 9.2.
The uniaxial stress and strain rate are denoted by σ and $\dot{e}$, respectively, and

$$\sigma = \eta \dot{e}, \tag{9.12}$$

where η is a viscosity coefficient. One-dimensional viscoelastic models are obtained by combining spring and dashpot elements. The two simplest models are the Maxwell fluid model that is represented schematically by a spring and dashpot connected in series and the Kelvin-Voigt solid model that is represented by a spring and dashpot in connected in parallel, shown in Figure 9.3.
The Maxwell model represents a viscoelastic fluid since it undergoes continuing deformation when a constant stress is applied. Referring to Figure 9.3a,

$$e = e_1 + e_2 = \frac{\sigma}{E} + e_2,$$

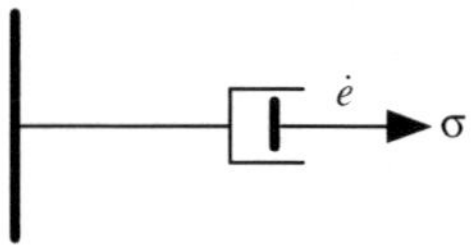

Figure 9.2. Viscous element.

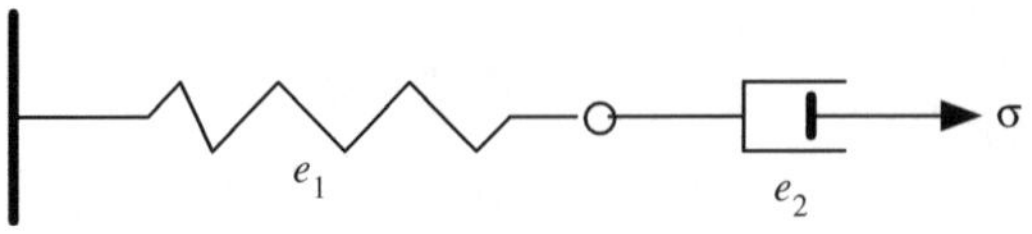

(a) Maxwell model.

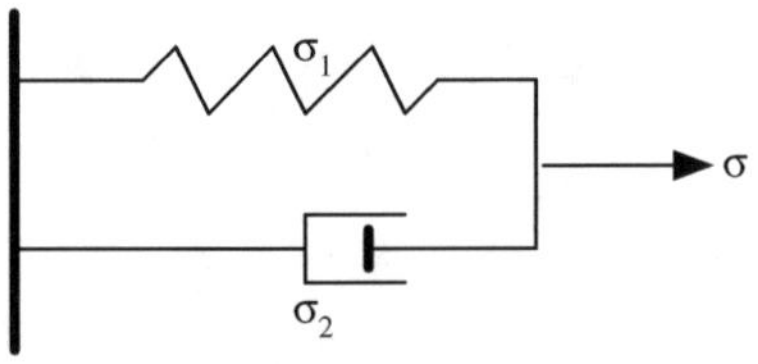

(b) Kelvin-Voigt model.

Figure 9.3. Newtonian viscous fluids. (a) Maxwell model. (b) Kelvin-Voigt model

and the time derivative and $\sigma = \eta \dot{e}_2$ give the constitutive equation for a Maxwell material,

$$\dot{\sigma} + \frac{E}{\eta}\sigma = E\dot{e}. \tag{9.13}$$

Suppose that at $t = 0^+, e = e_0$ and thereafter remains constant so that $\dot{e}(t) = 0$, that is, $e(t) = e_0 H(t)$, where $H(t)$ is the unit step function defined by

$$H(t) = \begin{cases} 1, t>0 \\ 0 \, t<0 \end{cases}.$$

It then follows from equation (9.13) and the initial condition $\sigma(0^+) = Ee_0$ that

$$\sigma(t) = Ee_0 \exp\left(-\frac{Et}{\eta}\right) H(t). \tag{9.14}$$

Equation (9.14) indicates relaxation, that is, the stress decays with time and

$$G(t) = E \exp\left(-\frac{Et}{\eta}\right) \tag{9.15}$$

is known as the relaxation function. The quantity η/E has dimension of time and is known as the relaxation time. In general the relaxation function represents the stress $\sigma(t)$ required to produce the strain $e(t) = e_0 H(t)$, that is,

$$\sigma(t) = G(t) e_0.$$

Since equation (9.14) indicates that $\sigma \to 0$, as $t \to \infty$, the relaxation modulus is zero. As $\dot{e} \to \infty$ the viscous element approaches rigidity, that is, $\dot{e}_2 \to 0$. Consequently, the impact modulus is E. Suppose now that the Maxwell model is subjected to a sudden application of axial stress $\sigma(t) = \sigma_0 H(t)$. For $t \geqslant 0^+$, $\dot{\sigma} = 0$ and $\sigma = \sigma_0$ and

$$\frac{E\sigma}{\eta} = E\dot{e},\ t \geqslant 0^+. \tag{9.16}$$

Then, it follows from equation (9.16) and initial condition $e(0^+)$ that

$$e = \sigma_0 \left\{ \frac{1}{E} + \frac{t}{\eta} \right\}. \tag{9.17}$$

Equation (9.17) indicates creep and that the Maxwell model represents a fluid since e increases monotonically due to a constant stress and

$$J(t) = \left\{ \frac{1}{E} + \frac{t}{\eta} \right\}. \tag{9.18}$$

is known as the creep function. In general the creep function is the strain $e(t)$ that results from the application of stress $\sigma(t) = \sigma_0 H(t)$, that is,

$$e(t) = J(t)\sigma_0.$$

The Maxwell model is now used to illustrate the superposition principle for linear viscoelasticity. Suppose a stress $\sigma(t)H(t)$ is applied to the Maxwell element, with $\sigma(0) = \sigma_*$, as indicated in Figure 9.4.

The response $de(t)$ due to stress increment $d\sigma$ applied at $t = \tau$,

$$de = d\sigma \left(\frac{1}{E} + \frac{(t - \tau)}{\eta} \right),\ t > \tau,$$

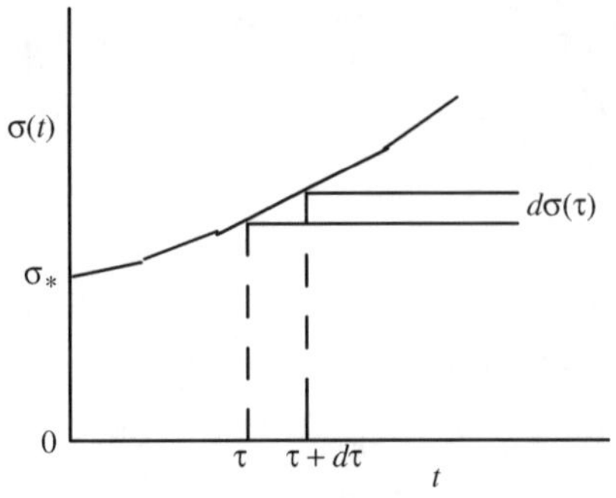

Figure 9.4. Application of stress increment.

follows from equation (9.17). Since the governing equation (9.13) is linear a superposition principle holds. Consequently,

$$e(t) = \sigma_* \left(\frac{1}{E} + \frac{t}{\eta}\right) + \int_{0^+}^{t} \frac{d\sigma(\tau)}{d\tau} \left(\frac{1}{E} + \frac{t-\tau}{\eta}\right) d\tau$$

or

$$e(t) = \int_{0^-}^{t} \frac{d\sigma(\tau)}{d\tau} \left(\frac{1}{E} + \frac{t-\tau}{\eta}\right) d\tau.$$

If $e(t)$ is prescribed the resulting stress $\sigma(t)$ is determined by solving equation (9.13).

Referring to Figure 9.3b for the Kelvin-Voigt model

$$\sigma = \sigma_1 + \sigma_2,$$

so that, with the use of equations (9.11) and (9.12), the constitutive equation is

$$\dot{e} + \frac{E}{\eta} e = \frac{\sigma}{\eta}. \tag{9.19}$$

Suppose the Kelvin-Voigt element is subjected to a sudden application of axial stress $\sigma(t) = H(t)\sigma_0$, then from equation (9.19),

$$\dot{e} + \frac{E}{\eta} e = \frac{\sigma_0}{\eta}. \tag{9.20}$$

The solution of equation (9.20) with initial condition $e(0) = 0$ is

$$e = \frac{\sigma_0}{E}\left\{1 - \exp\left(\frac{-Et}{\eta}\right)\right\}. \tag{9.21}$$

Equation (9.21) indicates creep since $e \to \sigma_0/E$ as $t \to \infty$. The creep function follows from equation (9.21) and is

$$J(t) = \frac{1}{E}\left\{1 - \exp\left(\frac{-Et}{\eta}\right)\right\}. \tag{9.22}$$

If at time t' the strain is fixed at constant value $e(t')$, σ immediately drops and remains constant with value

$$\sigma = Ee(t') = \sigma_0\left\{1 - \exp\left(\frac{-Et'}{\eta}\right)\right\},$$

that is, when the strain is held fixed, the stress immediately relaxes.

When the strain rate approaches infinity the viscous element approaches rigidity, consequently the impact modulus is infinite. Substituting $e(t) = e_0 H(t)$ in equation (9.19) gives

$$\sigma(t) = (\eta\delta(t) + E)e_0,$$

and the relaxation function is

$$G(t) = \eta\delta(t) + E, \tag{9.23}$$

where $\delta(t)$ is the Dirac delta function. The Dirac delta function $\delta(t)$, and the unit step function $H(t)$ belong to a class known as generalized functions. The unit function has already been defined and the Dirac delta function is defined by

$$\delta(t) = 0, t \neq 0, \text{ and } \int_{-\infty}^{\infty} \delta(t)dt = 1 \text{or} \int_{-\infty}^{\infty} \varphi(t)\delta(t)dt = \varphi(0),$$

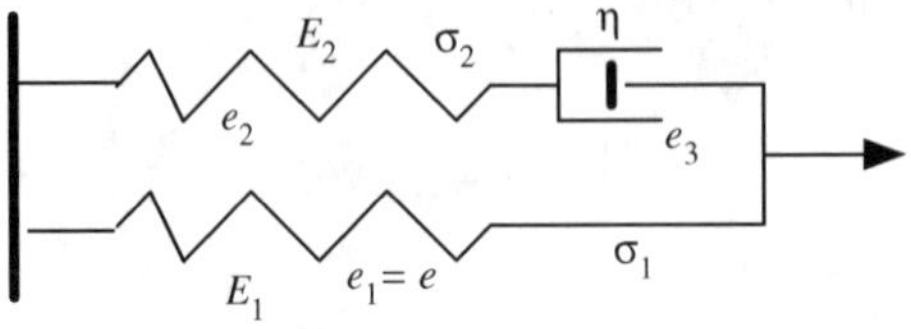

Figure 9.5. Standard model.

and can be regarded, in a nonrigorous way, as the time derivative of $H(t)$. The theory of generalized functions is given in a text by Lighthill [7]. It is evident from equation (9.23) that the Kelvin-Voigt model has an infinite impact modulus, and this indicates that it may not be suitable for certain applications.

There are many more complicated spring-dashpot models than the Maxwell and Kelvin-Voigt models. The so-called standard model consists of a spring and Maxwell element in parallel as indicated in Figure 9.5 and involves three parameters, two stiffness parameters, E_1 and E_2, and a viscosity η. It has been proposed as a more realistic model than the Kelvin-Voigt model for a linear viscoelastic solid. The standard model is used to illustrate the theory in the remainder of this chapter; however, the expressions for the hereditary integrals given are valid in general for isotropic linear viscoelasticity.

Referring to Figure 9.5,

$$e_2 + e_3 = e, \tag{9.24}$$

and

$$\sigma_1 + \sigma_2 = \sigma. \tag{9.25}$$

It follows from equations (9.11) and (9.12) and the time derivative of equation (9.24) that

$$\frac{\dot{\sigma}_2}{E_2} + \frac{\sigma_2}{\eta} = \dot{e}, \tag{9.26}$$

and from equations (9.25) and (9.11) and the time derivative of equation (9.25) that

$$\sigma_2 = \sigma - E_1 e, \tag{9.27}$$

$$\dot{\sigma}_2 = \dot{\sigma} - E_1 \dot{e}. \tag{9.28}$$

Elimination of σ_2 and $\dot{\sigma}_2$ from equations (9.26) to (9.28) gives the differential constitutive equation

$$\dot{\sigma} + \frac{E_2}{\eta}\sigma = (E_1 + E_2)\dot{e} + \frac{E_1 E_2}{\eta} e. \tag{9.29}$$

Equation (9.29) is equivalent to

$$\sigma = E_1 e + E_2(e - q) \tag{9.30}$$

and

$$\dot{q}(t) = \frac{E_2}{\eta}(e - q), \tag{9.31}$$

where $q = e_3$ is strain in the dashpot and is known as an internal variable. Equation (9.31) is known as the evolution equation for q.

Substitution of $e(t) = e_0 H(t)$ into equation (9.29) gives, for $t \geqslant 0^+$,

$$\dot{\sigma} + \frac{E_2}{\eta}\sigma = \frac{E_1 E_2}{\eta} e_0,$$

with solution

$$\sigma = \beta \exp\left(-\frac{E_2 t}{\eta}\right) + E_1 e_0, \tag{9.32}$$

where β is a constant of integration. At $t = 0^+$, the viscous element is instantaneously rigid and this gives the initial condition $\sigma(0^+) = (E_1 + E_2)e_0$ that in turn gives $\beta = E_2 e_0$ and equation (9.32) becomes

$$\sigma(t) = \left(E_2 \exp\left(-\frac{E_2 t}{\eta}\right) + E_1\right)e_0. \tag{9.33}$$

It then follows from equation (9.33) that the relaxation function for the standard model is

$$G(t) = \left(E_2 \exp\left(-\frac{E_2 t}{\eta}\right) + E_1\right), \tag{9.34}$$

$G(0) = E_1 + E_2$ is the impact modulus and $G(\infty) = E_1$ is the relaxation modulus.

The form of a typical relaxation function is shown in Figure 9.6. The impact modulus $G(0)$ for the Kelvin-Voigt model is infinite and the relaxation modulus $G(\infty)$ for the Maxwell model is zero.
Referring to Figure 9.7 and using superposition, a strain history $e(t)$, that has a step discontinuity e_* at $t = 0$ gives rise to

$$\sigma(t) = G(t)e_* + \int_{0^+}^{t} G(t-\tau)\frac{de(\tau)}{d\tau}d\tau, \tag{9.35a}$$

or

$$\sigma(t) = \int_{0^-}^{t} G(t-\tau)\frac{de(\tau)}{d\tau}d\tau. \tag{9.35b}$$

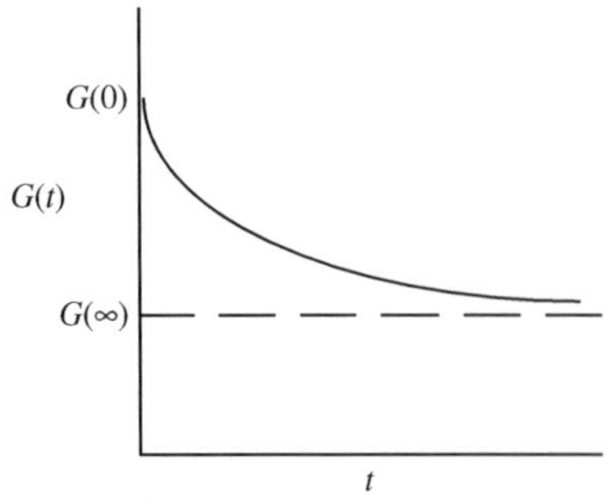

Figure 9.6. Typical relaxation function.

Equation (9.35) is described as a hereditary integral since $\sigma(t)$ depends on the strain history $e(\tau)$ for $\tau \leqslant t$.

In order to obtain the creep function for the standard model, $\sigma = \sigma_0 H(t)$ is substituted in equation (9.29). Then for $t \geqslant 0^+$,

$$\frac{E_2}{\eta}\sigma_0 = (E_1 + E_2)\dot{e} + \frac{E_1 E_2}{\eta} e,$$

with solution

$$e = \gamma \exp\left(\frac{-E_1 E_2 t}{\eta(E_1 + E_2)}\right) + \frac{\sigma_0}{E_1}, \tag{9.36}$$

where γ is a constant of integration. The initial condition is $e(0^+) = \sigma_0/(E_1 + E_2)$ and it follows from equation (9.36) that

$$e = \left\{\frac{1}{E_1} - \frac{E_2}{E_1(E_1 + E_2)}\exp\left(-\frac{E_1 E_2 t}{\eta(E_1 + E_2)}\right)\right\}\sigma_0.$$

and the creep function is

$$J(t) = \frac{1}{E_1} - \frac{E_2}{E_1(E_1 + E_2)}\exp\left(-\frac{E_1 E_2 t}{\eta(E_1 + E_2)}\right). \tag{9.37}$$

The form of a typical creep function is shown in Figure 9.8.

For the Maxwell model $J(t) \to \infty$ as $t \to \infty$ and for the Kelvin-Voigt model $J(0) = 0$.

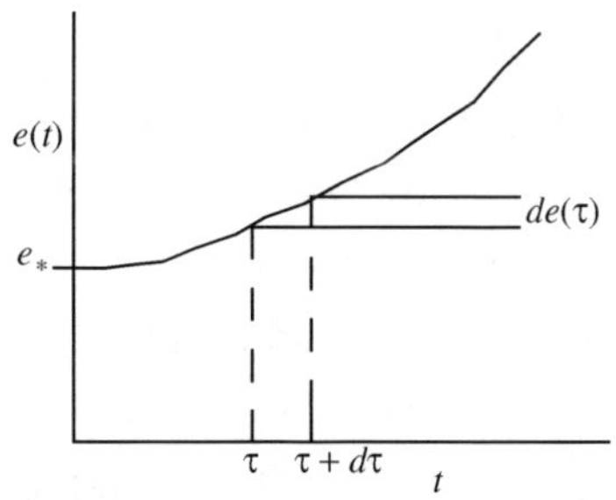

Figure 9.7. Application of strain increment.

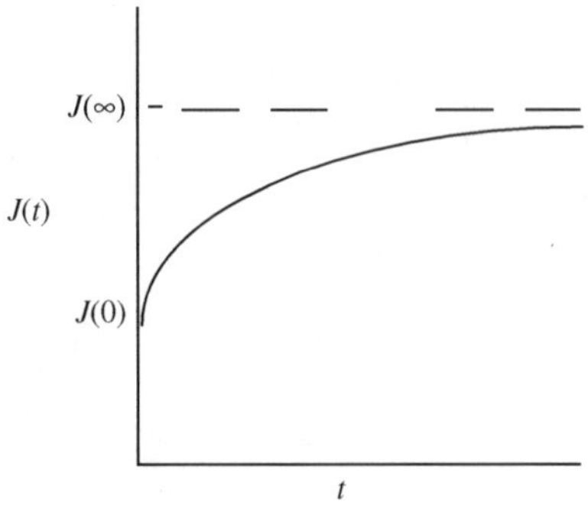

Figure 9.8. Typical creep function.

It follows from superposition that a strain history $\sigma(t)$ with a step discontinuity σ_* at $t = 0$ gives rise to another hereditary integral

$$e(t) = J(t)\sigma_* + \int_{0^+}^{t} J(t-\tau)\frac{d\sigma(\tau)}{d\tau}d\tau \tag{9.38}$$

or

$$e(t) = \int_{0^-}^{t} J(t-\tau)\frac{d\sigma(\tau)}{d\tau}d\tau. \tag{9.38b}$$

The forms (9.35) and (9.38), for the hereditary integrals, are valid for all discrete linear viscoelastic models and are valid for continuous systems.

A continuum is now considered and distortion and dilation relations are separated by expressing Cauchy stress, $\boldsymbol{\sigma}$, and infinitesimal strain, $\boldsymbol{e}$, in terms of the deviatoric and isotropic parts,

$$\sigma_{ij} = s_{ij} + \bar{\sigma}\delta_{ij},\ e_{ij} = e'_{ij} + \frac{\tilde{e}}{3}\delta_{ij},$$

where $\bar{\sigma} = \sigma_{kk}/3$ and $\tilde{e} = e_{kk}$.

The relaxation functions for distortion and dilatation described as the shear and bulk relaxation functions are denoted by $G_\mu(t)$ and $\mathrm{G}_K(t)$, respectively, and the corresponding creep functions are denoted by $J_\mu(t)$ and $J_K(t)$, respectively, and are obtained by replacing E_1 and E_2 in equations (9.34) and (9.37) by $2\mu_1$ and $2\mu_2$ for the distortional deformation and K_1 and K_2 for the dilatational deformation. Axial stress and strain in the relations already obtained for the one-dimensional viscoelastic model are now replaced by s_{ij} and e'_{ij} for distortional deformation and $\bar{\sigma}$ and $\tilde{e}$ for dilatation.

The hereditary integrals are

$$s_{ij} = \int_{0^-}^{t} G_\mu(t-\tau)\frac{de'_{ij}}{d\tau}d\tau, \tag{9.39}$$

$$\bar{\sigma} = \int_{0^-}^{t} G_K(t-\tau)\frac{d\tilde{e}}{d\tau}d\tau, \tag{9.40}$$

$$e'_{ij} = \int_{0^-}^{t} J_\mu(t-\tau)\frac{ds_{ij}}{d\tau}d\tau, \tag{9.41}$$

$$\tilde{e} = \int_{0^-}^{t} J_K(t-\tau)\frac{d\bar{\sigma}}{d\tau}d\tau. \tag{9.42}$$

The relaxation and creep functions G_K and J_K in equations (9.40) and (9.42), respectively, differ from those in [5] by a factor of 3, since they involve $\bar{\sigma} = \sigma_{kk}/3$ instead of σ_{kk}. The relaxation and creep functions in equations (9.39) to (9.42) for the continuum are of the same form as those for the one-dimensional discrete models, equations (9.34) and (9.38), with suitable choice of the parameters. For example, for the standard model each relaxation function involves three parameters and are of the form

$$G_\mu(t) = \left\{2\mu_2 \exp\left(-\frac{2\mu_2 t}{\eta}\right) + 2\mu_1\right\}, \tag{9.43}$$

with parameters $2\mu_1$, $2\mu_2$, and η, and

$$G_K(t) = \left\{K_2 \exp\left(-\frac{K_2 t}{\hat{\eta}}\right) + K_1\right\}, \tag{9.44}$$

with parameters K_1, K_2, and $\hat{\eta}$. The parameters μ_1 and K_1 corresponding to $G_\mu(\infty)$ and $G_K(\infty)$, respectively, are the relaxation shear and bulk moduli, respectively, and correspond to those of linear elasticity as expected.

Similarly, the corresponding impact moduli are $2(\mu_1 + \mu_2)$ and $(K_1 + K_2)$ and from equation (9.37) the creep functions for the continuum are

$$J_\mu = \frac{1}{2\mu_1} - \frac{2\mu_2}{4\mu_1(\mu_1 + \mu_2)} \exp\left(-\frac{4\mu_1\mu_2 t}{2\eta(\mu_1 + \mu_2)}\right), \tag{9.45}$$

$$J_K = \frac{1}{K_1} - \frac{K_2}{K_1(K_1 + K_2)} \exp\left(-\frac{K_1 K_2 t}{\hat{\eta}(K_1 + K_2)}\right). \tag{9.46}$$

It follows from equations (9.43) to (9.46) that

$$G_\mu(\infty)J_\mu(\infty) = 1 \text{ and } G_K(\infty)J_K(\infty) = 1 \tag{9.47}$$

and

$$G_\mu(0)J_\mu(0) = 1 \text{ and } G_K(0)J_K(0) = 1. \tag{9.48}$$

It should be noted that equations (9.47) and (9.48) do not imply that $G_\mu(t)J_\mu(t) = 1$ and $G_\mu(t)J_\mu(t) = 1$, for all t. The correct result [5] is obtained from

$$\bar{G}_\alpha(s)\bar{J}_\alpha(s) = s^{-2},$$

where the superposed bar denotes the Laplace transform and α is replaced by μ or K for shear or dilatational deformation, respectively. The Laplace transform $\bar{f}(s)$ of $f(t)$ is defined by

$$\bar{f}(s) = \int_0^\infty f(t)e^{-st}dt,$$

and a standard text on integral transforms should be consulted for details.

If it is assumed that the relaxation function for shear is proportional to that for dilatation, that is, $G_\mu(t)/G_K(t)$ is constant for all t. Then,

$$\frac{G_\mu(t)}{G_K(t)} = \frac{3(1-2\nu)}{1+\nu}$$

since $G_\mu(\infty)/G_K(\infty) = 2\mu_1/K_1$ and, for linear elasticity, $3K = 2\mu(1+v)/(1-2v)$, where v is Poisson's ratio.

Similarly, it may be shown that

$$\frac{J_\mu(t)}{J_K(t)} = \frac{1+v}{3(1-2v)}.$$

It may also be shown, for equations (9.43) and (9.44), that the relaxation times are equal, that is,

$$\frac{\eta}{2\mu_2} = \frac{\hat{\eta}}{K_2}$$

if $G_\mu(t)/G_K(t)$ is constant for all t.

9.4 Viscoelastic Dissipation

A continuum Kelvin-Voigt model, with the dilatation purely elastic, is perhaps the easiest linear viscoelastic model to obtain a relation for the rate of dissipation of mechanical energy. Referring to Figure 9.3b, the continuum form of the constitutive equation (9.19) is

$$\sigma_{ij} = 2\eta\dot{e}'_{ij} + 2\mu e_{ij} + \lambda e_{kk}\delta_{ij},$$

where $\lambda = K - 2\mu/3$, K and μ are the relaxation bulk and shear moduli, respectively. It then follows that the stress power per unit volume is

$$\sigma_{ij}\dot{e}_{ij} = 2\eta\dot{e}'_{ij}\dot{e}'_{ij} + 2\mu e_{ij}\dot{e}_{ij} + \lambda e_{kk}\dot{e}_{ll}.$$

The stress power is composed of two parts, a dissipative part, $2\eta\dot{e}'_{ij}\dot{e}'_{ij}$, and a part, $2\mu e_{ij}\dot{e}_{ij} + \lambda e_{kk}\dot{e}_{ll}$, which is the rate of change of stored elastic energy. It is reasonable to assume that, in general, the stress power is composed of a dissipative part and a rate of stored energy part. This is now shown for a more difficult case, a standard material. Continuum forms of the differential constitutive equation (9.29), for the shear and dilatational deformation, are

$$\dot{s}_{ij} + \frac{2\mu_2}{\eta}s_{ij} = 2(\mu_1 + \mu_2)\dot{e}'_{ij} + \frac{4\mu_1\mu_2}{\eta}e'_{ij} \tag{9.49}$$

and

$$\dot{\bar{\sigma}} + \frac{K_2}{\hat{\eta}}\bar{\sigma} = (K_1 + K_2)\dot{\tilde{e}} + \frac{K_1 K_2}{\hat{\eta}}\tilde{e}, \tag{9.50}$$

respectively. In equations (9.49) and (9.50) it is not assumed that the viscosity coefficient η is the same for shear and dilatation.

The stress power can be decomposed into a deviatoric or shear part, $P_\mu = s_{ij}\dot{e}'_{ij}$, and a dilatational part $P_K = \bar{\sigma}\dot{\tilde{e}}$ so that $\sigma_{ij}\dot{e}_{ij} = P_\mu + P_K$. A further decomposition of each of these parts into a dissipative part and a rate of stored energy part is possible. The part for the deviatoric deformation governed by equation (9.49) is now considered. Equation (9.49) is equivalent to [8],

$$s_{ij} = 2\mu_1 e'_{ij} + 2\mu_2 (e'_{ij} - q'_{ij}) \tag{9.51}$$

and

$$\dot{q}'_{ij} = \frac{2\mu_2}{\eta}(e'_{ij} - q'_{ij}). \tag{9.52}$$

Similarly, equation (9.50) is equivalent to

$$\bar{\sigma} = K_1\tilde{e} + K_2(\tilde{e} - \tilde{q}) \tag{9.53}$$

and

$$\dot{\tilde{q}} = \frac{K_2}{\hat{\eta}}(\tilde{e} - \tilde{q}), \tag{9.54}$$

where the tensor $\boldsymbol{q}$ with components $q_{ij} = q'_{ij} + (\tilde{q}/3)\delta_{ij}, \tilde{q} = q_{kk}$, is a so-called internal variable and is analogous to the strain of the dashpot element in Figure 9.5. More complicated viscoelastic materials whose spring and dashpot representations have more than one dashpot have a corresponding number of internal variables.

The stress power for the deviatoric part of the deformation obtained from equation (9.51) is

$$s_{ij}\dot{e}'_{ij} = 2\mu_1 e'_{ij}\dot{e}'_{ij} + 2\mu_2\left(e'_{ij} - q'_{ij}\right)\dot{e}'_{ij},$$

and with the use of equation (9.52) can be put in the form

$$s_{ij}\dot{e}'_{ij} = 2\mu_1 e'_{ij}\dot{e}'_{ij} + 2\mu_2\left(e'_{ij} - q'_{ij}\right)\left(\dot{e}'_{ij} - \dot{q}'_{ij}\right) + \eta\dot{q}'_{ij}\dot{q}'_{ij}. \tag{9.55}$$

If the parameters μ_1, μ_2, and η, and s_{ij} and e'_{ij} are known, q'_{ij} can be obtained from equation (9.51). The first two terms on the right-hand side of equation (9.55) represent the rate of storage of elastic strain energy and the third term represents the rate of dissipation of mechanical energy. This is evident if we consider the spring-dashpot model shown in Figure 9.5, since elastic energy is stored in the springs and mechanical energy is dissipated in the dashpot. Similarly, the part of the stress power for the dilatational deformation can be put in the form

$$\bar{\sigma}\tilde{e} = K_1\tilde{e}\dot{\tilde{e}} + K_2(\tilde{e} - \tilde{q})\left(\dot{\tilde{e}} - \dot{\tilde{q}}\right) + \hat{\eta}\dot{\tilde{q}}^2.$$

A model for linear viscoelasticity has been proposed [10], which assumes that the dilatational deformation is purely elastic, so that equations (9.40) and (9.42) are replaced by

$$\bar{\sigma} = K\tilde{e}.$$

All the relations already obtained for the deviatoric deformation of the standard model are valid, but for purely dilatational deformation linear elasticity theory holds. This model is somewhat similar to that for a compressible Newtonian viscous fluid with zero bulk viscosity and would be completely analogous for the Kelvin-Voigt viscoelastic model. A disadvantage of the model is that Poisson's ratio is not constant. It may be shown that for the standard model for deviatoric deformation along with purely elastic dilatation that the equilibrium and impact Poisson's ratios are given by

$$\nu_e = \frac{3K - 2\mu_1}{8K + 2\mu_1}$$

and

$$\nu = \frac{3K - 2(\mu_1 + \mu_2)}{8K + 2(\mu_1 + \mu_2)},$$

respectively.

The present treatment of linear viscoelasticity has so far neglected thermal effects, although the material properties may be temperature dependent. It is assumed that neglect of thermal effects may be justified if $|\Theta - \Theta_0|/\Theta_0 << 1$, where Θ_0 is the temperature in the undeformed reference configuration.

9.5 Some Thermodynamic Considerations in Linear Thermoelasticity

The linear theory of viscoelasticity, governed by equations (9.39) to (9.42), does not include thermal effects and may be regarded as isothermal. It also allows uncoupled decomposition into deviatoric and dilatational relations. Use of the theory implies the assumption of zero volume coefficient of thermal expansion and infinite thermal conductivity since mechanical dissipation produces temperature increase. It is clear that the usefulness of the linear theory is restricted to small temperature changes, that is, for $|\Theta - \Theta_0|/\Theta_0 << 1$, fortunately this condition is satisfied in many practical applications, for example, in structures.

The theory of thermoviscoelasticity has been developed in different ways by various authors [5]. Thurston [9] has shown that a difficulty arises when thermal effects are incorporated into linear viscoelasticity. According to Thurston that, in applying linear viscoelastictiy theory to adiabatic changes, the effect of energy dissipation gets linearized away. However, linear thermoviscoelastic theories have been presented by Christensen [5] and Hunter [10]. These theories will not be discussed here. In this section we give a short general outline of thermoviscoelastic theory, and in order to illustrate aspects of linear thermoviscoelasticity, we consider the standard model, with a purely elastic dilatational response, that has one internal variable with components q_{ij} such that $q_{kk} = 0$. The assumption of purely elastic response for dilatation enables a simple thermoviscoelastic theory to be developed.

In what follows specific thermodynamic problems are per unit mass. The energy equation (4.31) with d_{ij} replaced by $\dot{e}_{ij}$ and $u\, f + \Theta s$ is

$$\rho\big(\dot{f} + \dot{\Theta}s + \dot{s}\Theta\big) - \rho r - \sigma_{ij}\dot{e}_{ij} + \frac{\partial \mathrm{q_i}}{\partial x_i} = 0, \tag{9.56}$$

where q_i are the components of the heat flux vector, and the Clausius-Duhem in equality is

$$-\rho(s\dot{\Theta} + \dot{f}) + \sigma_{ij}\dot{e}_{ij} - \frac{\mathrm{q}_i}{\Theta}\frac{\partial \Theta}{\partial x_i} \geqslant 0. \tag{9.57}$$

It should be noted that equations (9.56) and (9.57) are valid only for infinitesimal deformation and $(\Theta - \Theta_0)/\Theta_0 << 1$. Consequently, it is reasonable to assume that the thermal conductivity k is constant and Fourier's law of heat conduction is given by

$$\mathrm{q}_i = -k\frac{\partial \Theta}{\partial x_i}.$$

In order to develop further a thermoviscoelastic theory σ_{ij} and e_{ij} are decomposed into their deviatoric and isotropic parts, $\sigma_{ij} = s_{ij} + \bar{\sigma}\delta_{ij}$ and $e_{ij} = e'_{ij} + (\tilde{e}/3)\delta_{ij}$.

The stress components are given by

$$\bar{\sigma} = K\Big(\tilde{e} - \alpha(\Theta - \Theta_0)\Big), \tag{9.58}$$

and

$$s_{ij} = 2\mu_1 e'_{ij} + 2\mu_2 \left(e'_{ij} - q_{ij}\right), \tag{9.59}$$

and the stress power by

$$\sigma_{ij}\dot{e}_{ij} = s_{ij}e'_{ij} + \bar{\sigma}\tilde{e},$$

where α is the coefficient of volume thermal expansion assumed constant for infinitesimal strain and $(\Theta - \Theta_0)/\Theta_0 << 1$. Equation (9.58) is the thermoelastic dilatation relation and equation (9.59) is identical to equation (9.51).

A possible form of the constitutive relation for the Helmholtz free energy is

$$f = f\left(e'_{ij}, \tilde{e}, \Theta, g_i, q_{ij}\right), \tag{9.60}$$

where $g_i = \partial\Theta/\partial x_i$ are the components of the temperature gradient vector. Substitution of

$$\dot{f} = \frac{\partial f}{\partial e'_{ij}}\dot{e}' + \frac{\partial f}{\partial \tilde{e}}\dot{\tilde{e}} + \frac{\partial f}{\partial \Theta}\dot{\Theta} + \frac{\partial f}{\partial g_i}\dot{g}_i + \frac{\partial f}{\partial q_{ij}}\dot{q}_{ij} \tag{9.61}$$

from equation (9.60) into equation (9.57) gives

$$\left(\frac{1}{\rho}s_{ij} - \frac{\partial f}{\partial e'_{ij}}\right)\dot{e}'_{ij} + \left(\frac{1}{\rho}\bar{\sigma} - \frac{\partial f}{\partial \tilde{e}}\right)\dot{\tilde{e}} - \left(s + \frac{\partial f}{\partial \Theta}\right)\dot{\Theta} \tag{9.62}$$

$$-\frac{\partial f}{\partial g_i}\dot{g}_i - \frac{\partial f}{\partial q_{ij}}\dot{q}_{ij} - \frac{1}{\rho\theta}\mathrm{q}_i g_i \geqslant 0.$$

Inequality equation (9.62) is valid for all values $\dot{e}'_{ij}$, $\dot{\tilde{e}}$, $\dot{\Theta}$, and $\dot{g}_i$, and this requires that the coefficients of $\dot{e}_{ij}$, $\dot{\Theta}$, and g_i must be zero. Consequently,

$$\sigma'_{ij} = \rho\frac{\partial f}{\partial e'_{ij}}, \tilde{\sigma} = \rho\frac{\partial f}{\partial \tilde{e}}, s = -\frac{\partial f}{\partial \theta}, \tag{9.63}$$

and f does not depend on g_i. Then, equations (9.62) and (9.58) become

$$-\frac{\partial f}{\partial q_{ij}}\dot{q}_{ij} - \frac{1}{\rho\theta}q_i g_i \geqslant 0, \tag{9.64}$$

$$f = f\left(e'_{ij}, \tilde{e}, \Theta, q_{ij}\right), \tag{9.65}$$

respectively.

The elimination of g_i from equation (9.60) is also justified by invoking the Principle of Local State. In equation (9.64), the first term is independent of g_i; consequently,

$$-\frac{\partial f}{\partial q_{ij}}\dot{q}_{ij} \geqslant 0, \tag{9.66}$$

and this is the rate of dissipation of mechanical energy. It follows from the mechanical dissipation term that appears in equation (9.55) and the expression (9.66) for mechanical dissipation that

$$-\rho\frac{\partial f}{\partial q_{ij}}\dot{q}_{ij} = \eta\dot{q}_{ij}\dot{q}_{ij}. \tag{9.67}$$

Then, from equations (9.52) and (9.67)

$$\rho\frac{\partial f}{\partial q_{ij}} = 2\mu_2\left(q_{ij} - e'_{ij}\right). \tag{9.68}$$

Substituting equations (9.63) and (9.68) into the differential of equation (9.65) and then substituting equations (9.58) and (9.59) gives

$$\begin{aligned} \rho df = {} & \left\{2(\mu_1+\mu_2)e'_{ij} - 2\mu_2 q_{ij}\right\}de'_{ij} + K\{\tilde{e} - \alpha(\Theta-\Theta_0)\}d\tilde{e} \\ & + \rho s d\Theta + 2\mu_2\left(q_{ij} - e'_{ij}\right)dq_{ij}. \end{aligned} \tag{9.69}$$

In order to integrate equation (9.69), it should be noted that $ds_{e'_{ij}=\tilde{e}=q_{ij=0}} = c_0 d\Theta/\Theta$ so that

$$s_{e'_{ij}=\tilde{e}=q_{ij=0}} = c_0\ln(\Theta/\Theta_0)$$

and linearizing gives

$$s_{e'_{ij}=\tilde{e}=q_{ij}=0}=c_0\left(\frac{\Theta-\Theta_0}{\Theta_0}\right), \tag{9.70}$$

where c_0 is the specific heat at constant deformation and is assumed constant. Using equation (9.70) it follows from equation (9.69) that

$$\begin{aligned}\rho f=&(\mu_1+\mu_2)e'_{ij}e'_{ij}-2\mu_2 q_{ij}e'_{ij}+K\frac{\tilde{e}^2}{2}\\&-K\alpha(\Theta-\Theta_0)\tilde{e}+\mu_2 q_{ij}q_{ij}-\frac{c_0}{2}\frac{(\Theta-\Theta_0)^2}{\Theta_0},\end{aligned} \tag{9.71}$$

and the linearized form of the entropy is

$$s=\frac{K\alpha\tilde{e}}{\rho}+c_0\left(\frac{\Theta-\Theta_0}{\Theta_0}\right),$$

which satisfies the relation

$$c_0=\Theta\frac{\partial s}{\partial\Theta}.$$

Substituting equations (9.63) and (9.68) in the time derivative of equation (9.65) or differentiating equation (9.71) with respect to time gives

$$\dot{f}=\frac{1}{\rho}s_{ij}\dot{e}'+\frac{1}{\rho}\bar{\sigma}\dot{\tilde{e}}-s\dot{\Theta}+2\mu_2\left(q_{ij}-e'_{ij}\right)\dot{q}_{ij}\leqslant\frac{1}{\rho}s_{ij}\dot{e}'+\frac{1}{\rho}\bar{\sigma}\dot{\tilde{e}}-s\dot{\Theta}, \tag{9.72}$$

where the inequality arises because of equation (9.66). Inequality equation (9.72) shows that the rate of change of the Helmhotz free energy is less than or equal to the stress power plus the "thermal power" minus $s\dot{\Theta}$ [8].

9.6 Simple Shear Problem

A problem of adiabatic simple shear involving the standard viscoelastic material with purely elastic dilatational response is now considered in order to illustrate some aspects of thermoelastic behavior. Simple shear is a homogeneous isochoric deformation and $e_{12} = e_{21}$ are the only nonzero components of the infinitesimal strain tensor and $q_{12} = q_{21}$ is the only nonzero components of the internal variable tensor. The problem considered is to determine the rate of dissipation of mechanical energy dissipation and the mechanical energy dissipated when equilibrium is attained if an application of strain

$$e_{12} = \varepsilon_{12} H(t) \tag{9.73}$$

is applied. Also the rate of dissipation of mechanical energy dissipation and the mechanical energy dissipated if unloading occurs from the loaded equilibrium state is determined.

It follows from equations (9.43) and (9.73) that

$$s_{12}(t) = \left\{ 2\mu_2 \exp\left(-\frac{2\mu_2 t}{\eta} \right) + 2\mu_1 \right\} \varepsilon_{12},$$

and from equations (9.51) and (9.52) that

$$s_{12} = 2\mu_2 e_{12} + \eta \dot{q}_{12}, \tag{9.74}$$

then

$$\dot{q}_{12} = \frac{\varepsilon_{12}}{T} \exp\left(\frac{-t}{T} \right). \tag{9.75}$$

The rate of mechanical energy dissipation per unit volume, obtained from equation (9.55), is given by

$$D = 2\eta q_{12}^2,$$

and with equation (9.75) this gives,

$$D = 2\eta \left[\frac{\varepsilon_{12}}{T} \exp\left(\frac{-t}{T} \right) \right]^2. \tag{9.76}$$

The mechanical energy dissipated, per unit volume when equilibrium has been reached is given by

$$W_d = 2\eta\left(\frac{\varepsilon_{12}}{T}\right)^2 \int_0^\infty \left[\exp\left(\frac{-t}{T}\right)\right]^2 dt = \eta\frac{\varepsilon_{12}^2}{T} = 2\mu_2\varepsilon_{12}^2, \qquad (9.77)$$

and this does not depend on the relaxation time.

An alternative derivation of the final result in equation (9.77) is as follows. When the strain (9.73) is applied, the impact modulus is $2(\mu_1 + \mu_2)$, and this results in an instantaneous stored elastic strain energy, $2(\mu_1 + \mu_2)\varepsilon_{12}^2$ per unit volume. When the equilibrium state is attained as $t \to \infty$, relaxation modulus is $2\mu_1$ and the stored elastic strain energy is $2\mu_1\varepsilon_{12}^2$. The difference between the stored elastic strain energies at $t(0)$ and $t(\infty)$ is $2\mu_2\varepsilon_{12}^2$ and this is in agreement with equation (9.77).

Now suppose the stress, $s_{12} = 2\mu_1 e_{12}$, at $t(\infty)$ is suddenly removed at time $\tilde{t} = 0$. A stress-free equilibrium state is then approached as $\tilde{t} \to \infty$ and the elastic stored energy is zero. Since the stored elastic strain energy at $\tilde{t} = 0$ is $2\mu_2\varepsilon_{12}^2$, it follows that this is the dissipation of mechanical energy for the unloading process.

EXERCISES

9.1. Show that equations (9.1) and (9.7) are frame indifferent.

9.2. Show that relation (9.10) is frame indifferent and $\text{tr}\boldsymbol{\sigma}/3 = -p$ if and only is the fluid is in equilibrium, that is, $\boldsymbol{D}\equiv\boldsymbol{0}$.

9.3. Determine the nonzero components of the extra stress $\boldsymbol{\sigma} + p\boldsymbol{I}$, given by equations (9.7) and (9.10) for simple shear flow with velocity field

$$v_1 = kx_2,\ v_2 = 0,\ v_3 = 0.$$

9.4. (a) Verify that $J(0) = (G(0))^{-1}$ for linear viscoelastic solids and fluids, and $J(\infty) = (G(\infty))^{-1}$ for linear viscoelastic solids.
(b) Show that $\bar{J} = \left(s^2\bar{G}\right)^{-1}$, where the superposed bar denotes the Laplace transform defined as $\bar{f}(s) = \int_0^\infty f(t)\exp(-st)dt$.

9.5. A Kelvin-Voigt element and a spring in series as shown in the figure is an alternative form of the three parameter model.

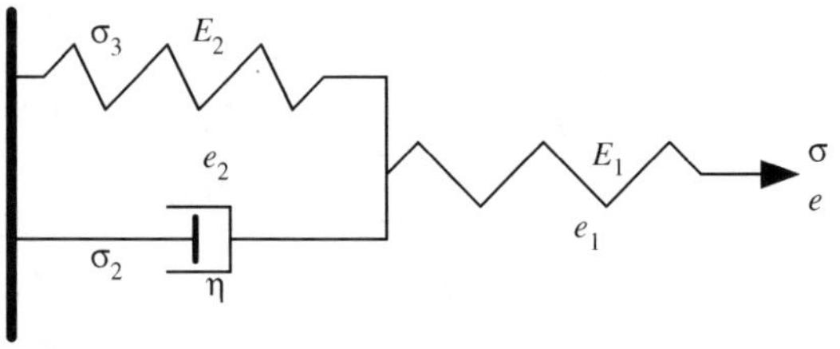

Exercise 5. Chapter 9.

Determine the differential constitutive relation, the relaxation function, and the creep function.

9.6. A spatially homogeneous simple shear deformation is given by

$$e_{12}(t) = 0, t<0,$$
$$e_{12}(t) = \varepsilon_{12}t/t', \quad 0 \leqslant t \leqslant t',$$
$$e_{12}(t) = \varepsilon_{12},$$

where ε_{12} is a constant strain. Show that for the standard material

$$s_{12} = 0, \quad t<0,$$

$$s_{12} = 2\mu_1 e_{12}(t) + 2\mu_2\varepsilon_{12}\frac{T}{t'}\left\{1 - \exp\left(\frac{-t}{T}\right)\right\}, 0 \leqslant t \leqslant t',$$

$$s_{12} = 2\mu_1\varepsilon_{12} + 2\mu_2\varepsilon_{12}\frac{T}{t'}\left\{\exp\left(\frac{t-t'}{T}\right) - \exp\left(\frac{-t}{T}\right)\right\}, t>t',$$

where $T = \eta/2\mu_2$ is the relaxation time.

9.7. A spring-Dashpot element shown in Figure 9.5 is subjected to a strain rate $\dot{e}_* H(t)$ where $\dot{e}_*$ is constant. Show that the stress-strain relation is given by

$$\sigma(e) = E_1 e + \eta\dot{e}_*\left\{1 - \exp\left(-\frac{E_2 e}{\eta\dot{e}_*}\right)\right\}.$$

9.8. A $(2N+1)$ parameter viscoelastic model, that is, a generalization of the three parameter standard model, is shown schematically in the figure.

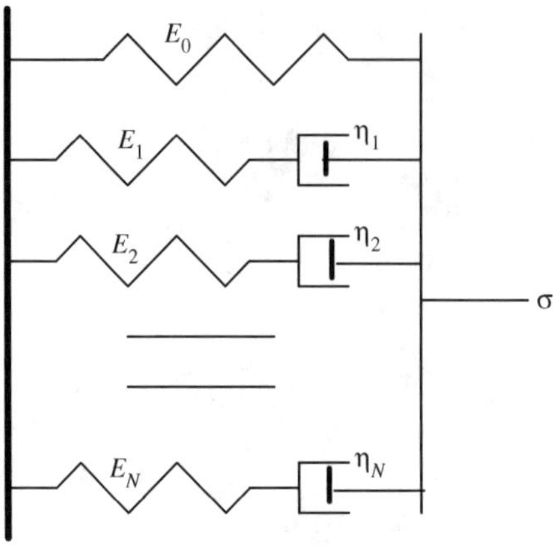

Exercise 8. Chapter 9.

Determine the set of relaxation times $T_k, k \in \{1, 2, ...N\}\tau$, the relaxation function and the creep function. The set of relaxation functions is known as the relaxation spectrum.

REFERENCES

1. Whitham, G.B. (1973). Linear and Nonlinear Waves. Wiley-Interscience.
2. Tanner, R.I. (1988). Engineering Rheology. Oxford University Press.
3. Schlichting, H. (1968). Boundary-Layer Theory, 6th Ed. McGraw-Hill.
4. Rivlin, R.S. (1948). The Hydrodynamics of Non-Newtonian Fluids I. Proc. Roy. Soc. London (A)193, pp. 260–281.
5. Christenen, R.M. (1982). Theory of Viscoelasicity, 2nd Ed. Academic Press.
6. Flügge, W. (1975). Viscoelasticity, 2nd Ed. Springer-Verlag.
7. Lighthill, M.J. (1958). An Introduction to Fourier Analysis and Generalised Functions. Cambridge University Press U.P.
8. Haupt, P. (2000). Continuum Mechanics and Theory of Materials. Springer.
9. Thurston, R.N. (1984). Waves in Solids. Mechanic of Solids, Vol. IV. Ed. C. Truesdell, Springer-Verlag.
10. Hunter, S.C. (1983). Mechanics of Continuous Media, 2nd Ed. Ellis Horwood.

Appendix 1

Orthogonal Curvilinear Coordinate Systems

A1.1 Introduction

Rectangular Cartesian coordinate systems are well suited for presenting the basic concepts of continuum mechanics since many mathematical complications that arise with other coordinate systems, for example, convected systems, are avoided. However, for the solution of specific problems, the use of rectangular Cartesian coordinate systems may result in considerable difficulties, and the use of other systems may be desirable in order to take advantage of symmetry aspects of the problem. For example, in the analysis of the expansion of a thick-walled cylindrical tube, the use of cylindrical polar coordinates has an obvious advantage. We introduce a simple theory of curvilinear coordinates in this appendix and specialize it for orthogonal curvilinear systems, in particular cylindrical and spherical. We avoid the use of the general theory of tensor components referred to curvilinear coordinates by considering what are known as the physical components of tensors that are derived for orthogonal coordinate systems. Superscripts are used to denote curvilinear coordinates.

A1.2 Curvilinear Coordinates

A point $\mathbf{y} \in E_3$ is the intersection of three mutually perpendicular planes whose equations are

$$x_1 = y_1, \quad x_2 = y_2, \quad x_3 = y_3$$

referred to the rectangular Cartesian coordinate system $0x_i$. Now introduce three scalar functions of $\boldsymbol{x}$,

$$\theta^j = \theta^j(x_1, x_2, x_3), \quad j \in \{1, 2, 3\}, \tag{A1.1}$$

with single valued inverses,

$$x_i = x_i\left(\theta^1, \theta^2, \theta^3\right). \tag{A1.2}$$

A superscript, rather than a suffix, is used for θ^j, since θ^j are not Cartesian coordinates. However, the summation convention applies as for suffixes. A point with rectangular coordinate x_i may also be specified by the values of θ^j at the point since equation (A1.2) represent three families of surfaces and the intersection of three surfaces $\theta^j = P^j$, where each P^j is a constant, locates a point. The P^j, $j \in \{1, 2, 3\}$ are the curvilinear coordinates of the point. It is assumed that the functions (A1.2) are at least of class C^1, that is, continuous with continuous first derivatives, and since equation (A1.2) must be invertible,

$$\det\left[\frac{\partial x_i}{\partial \theta^j}\right] \neq 0,$$

except at isolated points or on isolated lines. A set of functions (A1.1) is an admissible transformation of coordinates if these conditions are satisfied. The surfaces $\theta^j = P^j$ are called coordinate surfaces and the space curves formed by their intersections in pairs, as indicated in Figure 1.1A, are called coordinate curves.

Consider a point P with position vector $\boldsymbol{x}$, with respect to the origin of the rectangular Cartesian coordinate system $0x_i$. At P there are three vectors $\mathbf{g}_i$, $i \in \{1, 2, 3\}$ that are tangent to the coordinate curves and are given by

$$\mathbf{g}_i = \frac{\partial \boldsymbol{x}}{\partial \theta^i}. \tag{A1.3}$$

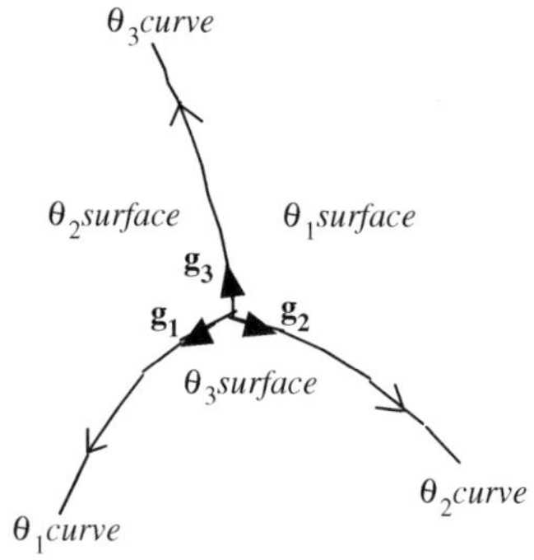

Figure 1.1A. Curvilinear coordinates.

The base vectors for rectangular Cartesian coordinate systems are given by $\mathbf{e}_j = \partial \boldsymbol{x}/\partial x_j$. Consequently,

$$\mathbf{g}_i = \mathbf{e}_s \frac{\partial x_s}{\partial \theta^i}, \quad \mathbf{e}_s = \mathbf{g}_i \frac{\partial \theta^i}{\partial x_s}. \tag{A1.4}$$

The vectors $\mathbf{g}_i$ are not in general unit vectors nor are they in general constants, except for the special case of oblique Cartesian coordinates, that is, when equations (A1.1) represent three nonorthogonal families of planes. Also the vectors $\mathbf{g}_i$ are nondimensional if and only if the corresponding θ^i have the dimension of length.

Referring to Figure 1.2A, the point P has position vector $\boldsymbol{x} = \boldsymbol{x}_i \mathbf{e}_i$ and curvilinear coordinates θ^j, and the neighboring point has position vector, $\boldsymbol{x} + d\boldsymbol{x} = (x_i + dx_i)\mathbf{e}_i$, and curvilinear coordinates $\theta^j + d\theta^j$.

If $\mathrm{PP}' = ds$,

$$ds^2 = d\boldsymbol{x} \cdot d\boldsymbol{x} = \mathbf{e}_i dx_i \cdot \mathbf{e}_j dx_j,$$

and using equation $(\mathrm{A}1.4)_2$, this becomes

$$ds^2 = \mathbf{g}_i \cdot \mathbf{g}_j d\theta^i d\theta^j = g_{ij} d\theta^i d\theta^j, \tag{A1.5}$$

where

$$g_{ij} = \mathbf{g}_i \cdot \mathbf{g}_j = \frac{\partial x_k}{\partial \theta^i} \frac{\partial x_k}{\partial \theta^j}, \tag{A1.6}$$

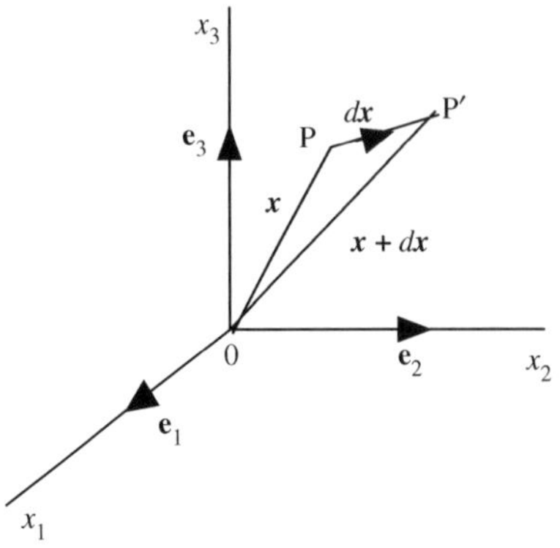

Figure 1.2A. Infinitesimal displacement of point P.

are the components of what is known as the metric tensor.

A1.3 Orthogonality

A curvilinear coordinate system is orthogonal if the vectors $\mathbf{g}_i$ form an orthogonal triad. It then follows from equation (A1.6) that

$$g_{ij} = 0 \text{ if } i \neq j,$$

and from equation (A1.5) that

$$ds^2 = g_{11}\left(d\theta^1\right)^2 + g_{22}\left(d\theta^2\right)^2 + g_{33}\left(d\theta^3\right)^2, \tag{A1.7}$$

where

$$g_{ii} = \left(\frac{\partial x_1}{\partial \theta^i}\right)^2 + \left(\frac{\partial x_2}{\partial \theta^i}\right)^2 + \left(\frac{\partial x_3}{\partial \theta^i}\right)^2, \text{(s.c.s)}. \tag{A1.8}$$

Henceforth (s.c.s), following an equation means that the summation convention is suspended and curvilinear coordinates only are considered. It is convenient to put

$$g_{11} = h_1^2, \quad g_{22} = h_2^2, \quad g_{33} = h_3^2. \tag{A1.9}$$

Unit base vectors $\mathbf{b}_i$ tangential to the coordinate curves at P are given by

$$\mathbf{b}_i = \frac{\mathbf{g}_i}{|\mathbf{g}_i|} = \frac{\mathbf{g}_i}{h_i}, \text{(s.c.s)}. \tag{A1.10}$$

The orthogonal triad of unit base vectors, $\mathbf{b}_i$, at P forms a basis for a localized Cartesian coordinate system; however, in general as P moves, the orientation of the triad changes. It follows that the vectors $\mathbf{b}_i$, are nondimensional and of constant unit magnitude but in general are not constant vectors.

Consider the vector field $\boldsymbol{u}(\boldsymbol{x})$ where $\boldsymbol{x}$ is the position vector of any point P as indicated in Figure 1.2A. The vector $\boldsymbol{u}(\boldsymbol{x})$ can be expressed in terms of its components with respect to the triad of unit base vectors $\mathbf{b}_i$, that is, with respect to the local Cartesian coordinate system at P as

$$\boldsymbol{u} = u_i\mathbf{b}_i.$$

The components u_i are known as the physical components of the vector and have the same physical dimension as $\boldsymbol{u}$. It should be noted that the vector must be associated with a particular point since two equal vectors associated with different points in space have in general different physical components since the unit base vectors, $\mathbf{b}_i$, are, in general, not constants.

A1.4 Cylindrical and Spherical Polar Coordinate Systems

The cylindrical polar coordinate system is the simplest orthogonal curvilinear coordinate system.

Referring to Figure 1.3A, P' is the orthogonal projection of P in the $0x_1x_2$ plane, and

$$\theta^1 = r = 0P', \quad \theta^2 = \phi = x_1\hat{0}P', \quad \theta^3 = z = PP'.$$

Consequently,

$$x_1 = r\cos\phi, \quad x_2 = r\sin\phi, \quad x_3 = z, \tag{A1.11}$$

where $r \geqslant 0$, $0 \leqslant \phi \leqslant 2\pi$, $-\infty < z < \infty$,

and these equations can be inverted to give

$$r = \left(x_1^2 + x_2^2\right)^{1/2}, \quad \phi = \arctan\frac{x_2}{x_1}, \quad z = x_3.$$

These equations represent a family of circular cylinders whose axes coincide with $0x_3$, a family of meridian planes and a family of planes normal to $0x_3$, respectively. It follows from equations (A1.7), (A1.8), and (A1.11) that

$$h_1 = 1, \quad h_2 = r, \quad h_3 = 1, \tag{A1.12}$$

and from equation (A1.6) that

$$ds^2 = dr^2 + r^2 d\phi^2 + dz^2.$$

The unit base vectors, $\mathbf{b}_1$, $\mathbf{b}_2$, and $\mathbf{b}_3$ are in the directions r, ϕ, and z increasing, respectively, as indicated in Figure 1.3A.

The spherical polar coordinates of P are indicated in Figure 1.4A.

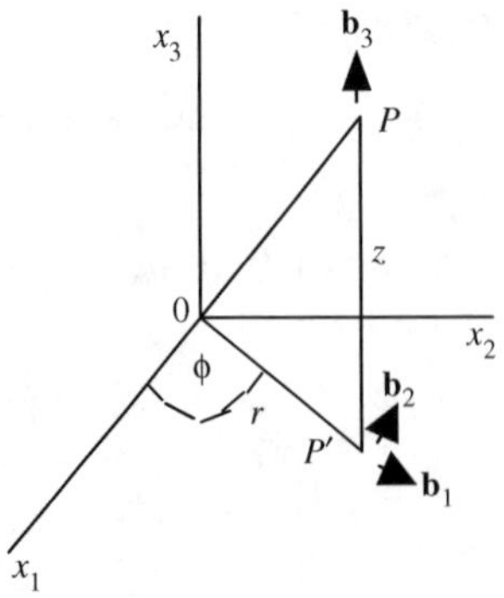

Figure 1.3A. Cylindrical polar coordinates.

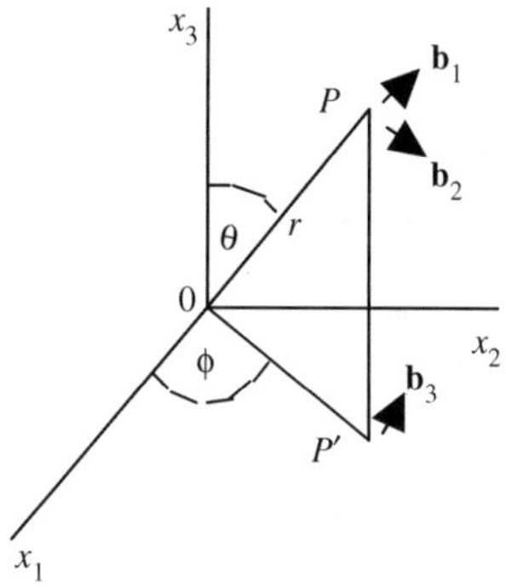

Figure 1.4A. Spherical polar coordinates.

Referring to Figure 1.4A, P' is the orthogonal projection of P in the $0x_1x_2$ plane, and

$$\theta^1 = r = 0P, \quad \theta^2 = \theta = x_3\hat{0}P, \quad \theta^3 = \phi = x_1 0P'.$$

Consequently,

$$x_1 = r\sin\theta\cos\phi, \quad x_2 = r\sin\theta\sin\phi, \quad x_3 = r\cos\phi, \tag{A1.13}$$

where $0 \leqslant \theta \leqslant \pi$, $0 \leqslant \phi \leqslant 2\pi$. The angles θ and ϕ are the colatitude and azimuth angles, respectively. Equation (A1.13) can be inverted to give

$$r = \left(x_1^2 + x_2^2 + x_3^2\right)^{1/2}, \quad \theta = \arctan\frac{\left(x_1^2 + x_2^2\right)}{x_3}, \quad \phi = \arctan\frac{x_2}{x_1},$$

which represent a family of spheres with centers at the origin, a family of cones with apexes at the origin, and a family of meridian planes, respectively. It follows from equations (A1.7), (A1.8), (A1.9), and (A1.13) that

$$h_1 = 1, \quad h_2 = r, \quad h_3 = r\sin\theta, \tag{A1.14}$$

and from equation (A1.6) that

$$ds^2 = dr^2 + r^2 d\theta^2 + r^2\sin^2\theta d\phi^2.$$

A1.5 The ∇ Operator

Rectangular Cartesian unit base vectors $\mathbf{e}_i$ are constants, consequently in applying either form of the gradient operator ∇, given by

$$\nabla \equiv \mathbf{e}_i \frac{\partial()}{\partial x_i}, \nabla \equiv \frac{\partial()}{\partial x_i} \mathbf{e}_i, \tag{A1.15}$$

to a nonscalar operand, say a vector or second-order tensor expressed in component form $\boldsymbol{u} = u_i \mathbf{e}_i$ or $\boldsymbol{S} = S_{ij} \mathbf{e}_i \otimes \mathbf{e}_j$, only the components are differentiated. This results in great simplification. At any point in a tensor field, a tensor can be expressed at that point in terms of its physical components with respect to an orthogonal curvilinear coordinate system with unit base vectors $\mathbf{b}_i$, for example,

$$\boldsymbol{S} = \tilde{S}_{ij} \mathbf{b}_i \otimes \mathbf{b}_j,$$

where $\tilde{S}_{ij}$ denotes the physical components of $\boldsymbol{S}$. Since the unit base vectors, $\mathbf{b}_i$, are not, in general, constants, this must be considered when $\boldsymbol{S}$ is operated on by a differential operator such as ∇ .

If Ψ is a scalar field, then at a point in the field

$$\nabla \Psi \cdot \mathbf{b}_i = \frac{d\Psi}{ds_i},$$

where $\mathbf{b}_i$ are the unit base vectors given by equation (A1.10) and s_i is the distance along the coordinate curve θ^i increasing.

Henceforth, the summation convention is suspended in this appendix.

It follows from equations (A1.7) and (A1.9) that

$$ds_i = h_i d\theta^i, \quad \in \{1, 2, 3\},$$

It may then be deduced that the orthogonal curvilinear form of equation $(A1.15)_2$ is

$$\nabla \equiv \sum_{i=1}^{3} \frac{1}{h_i} \frac{\partial()}{\partial \theta^i} \mathbf{b}_i. \tag{A1.16}$$

There is no difficulty in obtaining the gradient of a scalar field Ψ,

$$\nabla \Psi \equiv \sum_{i=1}^{3} \frac{1}{h_i} \frac{\partial \Psi}{\partial \theta^i} \mathbf{b}_i$$

since no base vectors have to be differentiated. The cylindrical polar form of the gradient operator obtained from equations (A1.12) and (A1.16) is

$$\nabla \equiv \frac{\partial()}{\partial r} \mathbf{b}_r + \frac{1}{r} \frac{\partial()}{\partial \phi} \mathbf{b}_\phi + \frac{\partial()}{\partial z} \mathbf{b}_z, \tag{A1.17}$$

where $\mathbf{b}_1$, $\mathbf{b}_2$, and $\mathbf{b}_3$ are replaced by $\mathbf{b}_r$, $\mathbf{b}_\phi$, and $\mathbf{b}_z$, and the spherical polar form is

$$\nabla \equiv \frac{\partial()}{\partial r} \mathbf{b}_r + \frac{1}{r} \frac{\partial()}{\partial \theta} \mathbf{b}_\theta + \frac{1}{r \sin \theta} \frac{\partial()}{\partial \phi} \mathbf{b}_\phi. \tag{A1.18}$$

where $\mathbf{b}_1, \mathbf{b}_2$, and $\mathbf{b}_3$ are replaced by $\mathbf{b}_r, \mathbf{b}_\theta$, and $\mathbf{b}_\phi$.

In order to obtain expressions for $\mathrm{div}\boldsymbol{u}, \mathrm{grad}\boldsymbol{u}, \mathrm{curl}\boldsymbol{u}, \mathrm{div}\boldsymbol{S}$, etc., in terms of the physical components of an orthogonal curvilinear coordinate system the elements of the matrix

$$\left[\frac{\partial \mathbf{b}_i}{\partial \theta^j}\right] \tag{A1.19}$$

are required. It follows from equations (A1.3) and (A1.10) that

$$\frac{\partial^2 \boldsymbol{x}}{\partial \theta^i \partial \theta^j} = \frac{\partial (h_j \mathbf{b}_j)}{\partial \theta^i} = \frac{\partial (h_i \mathbf{b}_i)}{\partial \theta^j}, i \neq j, \text{(s.c.s)}$$

and differentiating gives

$$h_j \frac{\partial \mathbf{b}_j}{\partial \theta^i} + \frac{\partial h_j}{\partial \theta^i} \mathbf{b}_j = h_i \frac{\partial \mathbf{b}_i}{\partial \theta^j} + \frac{\partial h_i}{\partial \theta^j} \mathbf{b}_i. \tag{A1.20}$$

Since the corresponding components of equal vectors are equal it follows from equation (A1.20) that

$$\frac{\partial \mathbf{b}_i}{\partial \theta^j} = \frac{1}{h_i} \frac{\partial h_j}{\partial \theta^i} \mathbf{b}_j, \tag{A1.21}$$

for the off diagonal elements of equation (A1.19).

The diagonal elements of equation (A1.19),

$$\frac{\partial \mathbf{b}_i}{\partial \theta^i} = -\frac{1}{h_j} \frac{\partial h_i}{\partial \theta^j} \mathbf{b}_j - \frac{1}{h_k} \frac{\partial h_i}{\partial \theta^k} \mathbf{b}_k \tag{A1.22}$$

are obtained from

$$\frac{\partial \mathbf{b}_i}{\partial \theta^i} = \frac{\partial \left(\mathbf{b}_j \times \mathbf{b}_k \right)}{\partial \theta^i}$$

and equation (A1.20). Equations (A1.21) and (A1.22) give

$$\left[\frac{\partial \mathbf{b}_i}{\partial \theta^j} \right] = \begin{bmatrix} 0 & \mathbf{b}_\phi & 0 \\ 0 & -\mathbf{b}_r & 0 \\ 0 & 0 & 0 \end{bmatrix} \tag{A1.23}$$

for cylindrical polar coordinates, and

$$\left[\frac{\partial \mathbf{b}_i}{\partial \theta^j} \right] = \begin{bmatrix} 0 & \mathbf{b}_\theta & \mathbf{b}_\phi \sin\theta \\ 0 & -\mathbf{b}_r & \mathbf{b}_\phi \cos\theta \\ 0 & 0 & -\mathbf{b}_r \sin\theta - \mathbf{b}_\theta \cos\theta \end{bmatrix} \tag{A1.24}$$

for spherical polar coordinates.

The divergence of a vector $\boldsymbol{u}$ referred to orthogonal curvilinear coordinates,

$$\mathrm{div}\boldsymbol{u} = \sum_{i=1}^{3} \frac{1}{h_i} \frac{\partial u}{\partial \theta^i} \cdot \mathbf{b}_i,$$

is obtained from equation (A1.16) and substitution of $\boldsymbol{u} = \mathbf{b}_1 u_1 + \mathbf{b}_2 u_2 + \mathbf{b}_3 u_3$ gives

$$\mathrm{div}\boldsymbol{u} = \frac{1}{h_1 h_2 h_3} \left[\frac{\partial(u_1 h_2 h_3)}{\partial \theta^1} + \frac{\partial(u_2 h_1 h_3)}{\partial \theta^2} + \frac{\partial(u_3 h_1 h_2)}{\partial \theta^3} \right]. \tag{A1.25}$$

For cylindrical polar coordinates

$$\mathrm{div}\boldsymbol{u} = \frac{\partial u_r}{\partial r} + \frac{u_r}{r} + \frac{1}{r}\frac{\partial u_\phi}{\partial \phi} + \frac{\partial u_z}{\partial z}$$

is obtained from equations (A1.12) and (A1.25), and for spherical polar coordinates

$$\mathrm{div}\boldsymbol{u} = \frac{1}{r^2}\frac{\partial\left(u_r r^2\right)}{\partial r} + \frac{1}{r \sin\theta}\frac{\partial(u_\theta \sin\theta)}{\partial \theta} + \frac{1}{r\sin\theta}\frac{\partial u_\phi}{\partial \phi}$$

is obtained from equations (A1.14) and (A1.25).

The definition of the curl of a vector $\boldsymbol{u}$ is usually based on the form equation $(A1.15)_1$ of the gradient operator consequently,

$$\mathrm{curl}\boldsymbol{u} = -\nabla \times \boldsymbol{u},$$

where ∇ is of the form $(A1.15)_2$. This is because of the anticommutative property of the vector product. It then follows that, for orthogonal curvilinear coordinates

$$\mathrm{curl}\boldsymbol{u} = -\sum_{i=1}^{3} \frac{1}{h_i} \frac{\partial u}{\partial \theta^i} \times \mathbf{b}_i,$$

and, after substitution of $\boldsymbol{u} = \mathbf{b}_1 u_1 + \mathbf{b}_2 u_2 + \mathbf{b}_3 u_3$,

$$\operatorname{curl}\boldsymbol{u} = \frac{1}{h_1 h_2 h_3} \det \begin{bmatrix} h_1\mathbf{b}_1 & h_2\mathbf{b}_2 & h_3\mathbf{b}_3 \\ \dfrac{\partial}{\partial\theta^1} & \dfrac{\partial}{\partial\theta^2} & \dfrac{\partial}{\partial\theta^3} \\ h_1 u_1 & h_2 u_2 & h_3 u_3 \end{bmatrix}. \tag{A1.26}$$

For cylindrical polar coordinates

$$\operatorname{curl}\boldsymbol{u} = \mathbf{b}_r\left(\frac{1}{r}\frac{\partial u_z}{\partial\phi} - \frac{\partial u_\phi}{\partial z}\right) + \mathbf{b}_\phi\left(\frac{\partial u_r}{\partial z} - \frac{\partial u_z}{\partial z}\right) + \mathbf{b}_z\left(\frac{1}{r}\frac{\partial(ru_\phi)}{\partial r} - \frac{1}{r}\frac{\partial u_z}{\partial\phi}\right)$$

is obtained from equations (A1.12) and (A1.265). For spherical polar coordinates

$$\operatorname{curl}\boldsymbol{u} = \frac{\mathbf{b}_r}{r\sin\theta}\left(\frac{\partial(u_\phi\sin\theta)}{\partial\theta} - \frac{\partial u_\theta}{\partial\phi}\right) + \frac{\mathbf{b}_\theta}{r}\left(\frac{1}{\sin\theta}\frac{\partial u_r}{\partial\phi} - \frac{\partial(ru_\phi)}{\partial r}\right)$$
$$+ \frac{\mathbf{b}_\phi}{r}\left(\frac{\partial(ru_\theta)}{\partial r} - \frac{\partial u_r}{\partial\theta}\right)$$

is obtained from equations (A1.14) and (A1.26).

The gradient of the vector $\boldsymbol{u}$ has in general nine components and is given by

$$\nabla\otimes\boldsymbol{u} = \sum_{i=1}^{3}\frac{1}{h_i}\frac{\partial\boldsymbol{u}}{\partial\theta^i}\otimes\mathbf{b}_i, \tag{A1.27}$$

where a second summation arises when $\partial\boldsymbol{u}/\partial\theta^i$ is evaluated using equations (A1.21) and (A1.22). It follows from equations (A1.23) and (A1.27) that, for cylindrical polar coordinates, the matrix of components of $\nabla\otimes\boldsymbol{u}$ is

$$[\nabla\otimes\boldsymbol{u}] = \begin{bmatrix} \dfrac{\partial u_r}{\partial r} & \dfrac{1}{r}\left(\dfrac{\partial u_r}{\partial\phi} - a_\phi\right) & \dfrac{\partial u_r}{\partial z} \\ \dfrac{\partial u_\phi}{\partial r} & \dfrac{1}{r}\left(\dfrac{\partial u_\phi}{\partial\phi} - a_\phi\right) & \dfrac{\partial u_\phi}{\partial z} \\ \dfrac{\partial u_z}{\partial r} & \dfrac{1}{r}\dfrac{\partial u_z}{\partial\phi} & \dfrac{\partial u_z}{\partial z} \end{bmatrix},$$

and from equations (A1.24) and (A1.27) that, for spherical polar coordinates,

$$[\nabla \otimes \boldsymbol{u}] = \begin{bmatrix} \frac{\partial u_R}{\partial r} & \frac{1}{r}\left(\frac{\partial u_r}{\partial \theta} - u_\theta\right) & \frac{1}{r\sin\theta}\left(\frac{\partial u_r}{\partial \phi} - u_\phi \sin\theta\right) \\ \frac{\partial u_\theta}{\partial r} & \frac{1}{r}\left(\frac{\partial u_\theta}{\partial \theta} + u_R\right) & \frac{1}{r\sin\theta}\left(\frac{\partial u_\theta}{\partial \phi} - u_\phi \cos\theta\right) \\ \frac{\partial u_\phi}{\partial r} & \frac{1}{r}\frac{\partial u_\phi}{\partial \theta} & \frac{1}{r\sin\theta}\left(\frac{\partial u_\phi}{\partial \phi} + u_\theta \cos\theta + u_r \sin\theta\right) \end{bmatrix}.$$

A1.6 The ∇^2 Operator

The ∇^2 operator, sometimes known as the Laplacian operator, is a scalar differential operator given by

$$\nabla^2 \equiv \sum_{i=1}^{3} \frac{\partial^2}{\partial x_i \partial x_i}$$

referred to a rectangular Cartesian coordinate system. This operator, which may operate on a scalar, vector, or higher-order tensor can be expressed as

$$\nabla^2 \equiv \text{div grad} = \nabla \cdot \nabla,$$

and with the use of equation (A1.16),

$$\nabla^2 \equiv \frac{1}{h_1 h_2 h_3}\left\{\frac{\partial}{\partial \theta^1}\left(\frac{h_2 h_3}{h_1}\frac{\partial()}{\partial \theta^1}\right) + \frac{\partial}{\partial \theta^2}\left(\frac{h_3 h_1}{h_2}\frac{\partial()}{\partial \theta^2}\right) + \frac{\partial}{\partial \theta^3}\left(\frac{h_1 h_2}{h_3}\frac{\partial()}{\partial \theta^3}\right)\right\}, \tag{A1.28}$$

referred to orthogonal curvilinear coordinates. For cylindrical polar coordinates, it follows from equation (A1.28) that

$$\nabla^2 \equiv \frac{\partial^2}{\partial r^2} + \frac{1}{r}\frac{\partial}{\partial r} + \frac{1}{r^2}\frac{\partial^2}{\partial \phi^2} + \frac{\partial^2}{\partial z^2}. \tag{A1.29}$$

and for spherical polar coordinates

$$\nabla^2 \equiv \frac{1}{r^2}\frac{\partial}{\partial r}\left(r^2\frac{\partial()}{\partial r}\right) + \frac{1}{r^2 \sin\theta}\frac{\partial}{\partial \theta}\left(\sin\theta\frac{\partial()}{\partial \theta}\right) + \frac{1}{r^2\sin^2\theta}\frac{\partial^2()}{\partial \phi^2}. \quad \text{(A1.30)}$$

When equation (A1.29) or (A1.30) operates on a vector it is important to note that the base vectors are not, in general, constants, which results in some complication, as indicated in this example, for cylindrical polar coordinates,

$$\nabla^2 \boldsymbol{u} = \mathbf{b}_r\left(\nabla^2 u_r - \frac{u_r}{r^2} - \frac{2}{r^2}\frac{\partial u_\phi}{\partial \phi}\right) + \mathbf{b}_\phi\left(\nabla^2 u_\phi - \frac{u_\phi}{r^2} + \frac{2}{r^2}\frac{\partial u_r}{\partial \phi}\right) + \mathbf{b}_z \nabla^2 u_z.$$

The terms that result from the nonconstant base vectors, $\mathbf{b}_r$ and $\mathbf{b}_\phi$, are clearly evident.

Appendix 2

Physical Components of the Deformation Gradient Tensor

The concept of physical components of vectors and second-order tensors, with respect to orthogonal curvilinear coordinate systems, was introduced in Appendix 1. In this appendix, extension to two-point tensors is considered, in particular, the deformation gradient tensor,

$$\boldsymbol{F} = \nabla_X \boldsymbol{x},$$

given in terms of its Cartesian components by

$$\boldsymbol{F} = \frac{\partial x_i}{\partial X_K} \mathbf{e}_i \otimes \mathbf{E}_K,$$

in the notation of chapter 2. In what follows lowercase letters refer to the spatial configuration and uppercase letters to the natural reference configuration as for the rectangular Cartesian expressions. The curvilinear coordinates and base vectors, referred to the reference configuration, are (R, Φ, Z) and $(\boldsymbol{B_R}, \boldsymbol{B}_\Phi, \boldsymbol{B}_Z)$, respectively, for cylindrical polar coordinates, and (R, Θ, Φ) and $(\boldsymbol{B}_R, \boldsymbol{B}_\Theta, \boldsymbol{B}_\Phi)$, respectively, for spherical polar coordinates. Then, the gradient operator ∇_X, referred to cylindrical polar coordinates, is

$$\nabla_X \equiv \frac{\partial()}{\partial R}\boldsymbol{B}_R + \frac{1}{R}\frac{\partial()}{\partial \Phi}\boldsymbol{B}_\Phi + \frac{\partial()}{\partial Z}\boldsymbol{B}_Z, \tag{A2.1}$$

and referred to spherical polar coordinates is

$$\nabla_X \equiv \frac{\partial()}{\partial R}\boldsymbol{B}_R + \frac{1}{R}\frac{\partial()}{\partial \Theta}\boldsymbol{B}_\Theta + \frac{1}{R\sin\Theta}\frac{\partial()}{\partial \Phi}\boldsymbol{B}_\Phi \tag{A2.2}$$

These are of the same form as the corresponding forms of ∇ given in Appendix 1.

The position vector $\boldsymbol{x}$ of a material particle in the spatial configuration is given, in terms of its cylindrical polar coordinates (r, ϕ, z), by

$$\boldsymbol{x} = r\mathbf{b}_r + z\mathbf{b}_z, \tag{A2.3}$$

and in terms of its spherical polar coordinates (r, θ, ϕ), by

$$\boldsymbol{x} = r\mathbf{b}_r. \tag{A.2.4}$$

The component matrix of $\boldsymbol{F}$ for cylindrical polar coordinates is now obtained. It follows from equations (A2.1) and (A2.3) that $\boldsymbol{F}$ can be expressed in the form

$$\boldsymbol{F} = \frac{\partial}{\partial R}(r\mathbf{b}_r + z\mathbf{b}_z) \otimes \boldsymbol{B}_R + \frac{1}{R}\frac{\partial}{\partial \Phi}(r\mathbf{b}_r + z\mathbf{b}_z) \otimes \boldsymbol{B}_\Phi + \frac{\partial}{\partial Z}(r\mathbf{b}_r + z\mathbf{b}_z) \otimes \mathbf{B}_Z. \tag{A2.5}$$

The matrix (A1.22),

$$\left[\frac{\partial b_i}{\partial \theta^j}\right] = \begin{bmatrix} 0 & \mathbf{b}_\phi & 0 \\ 0 & -\mathbf{b}_r & 0 \\ 0 & 0 & 0 \end{bmatrix}, \; i, j \in \{r, \phi, z\},$$

is required in order to perform the differentiations in equation (A2.5). Then

$$\frac{\partial}{\partial R}\begin{bmatrix} \mathbf{b}_r \\ \mathbf{b}_\phi \\ \mathbf{b}_z \end{bmatrix} = \frac{\partial}{\partial r}\begin{bmatrix} \mathbf{b}_r \\ \mathbf{b}_\phi \\ \mathbf{b}_z \end{bmatrix}\frac{\partial r}{\partial R} + \frac{\partial}{\partial \phi}\begin{bmatrix} \mathbf{b}_r \\ \mathbf{b}_\phi \\ \mathbf{b}_z \end{bmatrix}\frac{\partial \phi}{\partial R} + \frac{\partial}{\partial z}\begin{bmatrix} \mathbf{b}_r \\ \mathbf{b}_\phi \\ \mathbf{b}_z \end{bmatrix}\frac{\partial z}{\partial R} = \begin{bmatrix} \mathbf{b}_\phi \\ -\mathbf{b}_r \\ \mathbf{0} \end{bmatrix}\frac{\partial \phi}{\partial R}. \tag{A2.6}$$

Similarly

$$\frac{\partial}{\partial \Phi}\begin{bmatrix}\mathbf{b}_r\\ \mathbf{b}_\phi\\ \mathbf{b}_z\end{bmatrix} = \begin{bmatrix}\mathbf{b}_\phi\\ -\mathbf{b}_r\\ \mathbf{0}\end{bmatrix}\frac{\partial \phi}{\partial \Phi} \tag{A2.7}$$

and

$$\frac{\partial}{\partial Z}\begin{bmatrix}\mathbf{b}_r\\ \mathbf{b}_\phi\\ \mathbf{b}_z\end{bmatrix} = \begin{bmatrix}\mathbf{b}_\phi\\ -\mathbf{b}_r\\ \mathbf{0}\end{bmatrix}\frac{\partial \phi}{\partial Z}. \tag{A2.8}$$

Use of relations (A2.6) to (A2.8) along with equation (A2.3) gives

$$\begin{aligned}\boldsymbol{F} = {}& \frac{\partial r}{\partial R}\mathbf{b}_r \otimes \boldsymbol{B}_R + \frac{1}{R}\frac{\partial r}{\partial \Phi}\mathbf{b}_r \otimes \boldsymbol{B}_\Phi + \frac{\partial r}{\partial Z}\mathbf{b}_r \otimes \boldsymbol{B}_Z +\\ & r\frac{\partial \phi}{\partial R}\mathbf{b}_\phi \otimes \boldsymbol{B}_R + \frac{r}{R}\frac{\partial \phi}{\partial \Phi}\mathbf{b}_\phi \otimes \boldsymbol{B}_\Phi + r\frac{\partial \phi}{\partial Z}\mathbf{b}_\phi \otimes \boldsymbol{B}_Z +\\ & \frac{\partial z}{\partial R}\mathbf{b}_z \otimes \boldsymbol{B}_R + \frac{1}{R}\frac{\partial z}{\partial \Phi}\mathbf{b}_z \otimes \boldsymbol{B}_\Phi + \frac{\partial z}{\partial Z}\mathbf{b}_z \otimes \boldsymbol{B}_Z.\end{aligned}$$

Consequently,

$$\begin{bmatrix}F_{rR} & F_{r\Phi} & F_{rZ}\\ F_{\phi R} & F_{\phi\Phi} & F_{\phi Z}\\ F_{zR} & F_{z\Phi} & F_{zZ}\end{bmatrix} = \begin{bmatrix}\frac{\partial r}{\partial R} & \frac{1}{R}\frac{\partial r}{\partial \Phi} & \frac{\partial r}{\partial Z}\\ r\frac{\partial \phi}{\partial R} & \frac{r}{R}\frac{\partial \phi}{\partial \Phi} & r\frac{\partial \phi}{\partial Z}\\ \frac{\partial z}{\partial R} & \frac{1}{R}\frac{\partial z}{\partial \Phi} & \frac{\partial z}{\partial Z}\end{bmatrix}. \tag{A2.7}$$

The component matrix of **F** for spherical polar coordinates

$$\begin{bmatrix}F_{rR} & F_{r\Theta} & F_{r\Phi}\\ F_{\theta R} & F_{\theta\Theta} & F_{\theta\Phi}\\ F_{\phi R} & F_{\phi\Theta} & F_{\phi\Phi}\end{bmatrix} = \begin{bmatrix}\frac{\partial r}{\partial R} & \frac{1}{R}\frac{\partial r}{\partial \Theta} & \frac{1}{R\sin\theta}\frac{\partial r}{\partial \Phi}\\ r\frac{\partial \theta}{\partial R} & \frac{r}{R}\frac{\partial \theta}{\partial \Theta} & \frac{r}{R\sin\theta}\frac{\partial \theta}{\partial \Phi}\\ r\sin\theta\frac{\partial \phi}{\partial R} & \frac{r}{R}\sin\theta\frac{\partial \phi}{\partial \Theta} & \frac{r}{R}\frac{\partial \phi}{\partial \Phi}\end{bmatrix} \tag{A2.8}$$

is obtained in a similar manner.

The physical components of other tensors such $\boldsymbol{B} = \boldsymbol{F}\boldsymbol{F}^T$, $C = \boldsymbol{F}^T\boldsymbol{F}$, and $\boldsymbol{E} = \frac{1}{2}(\boldsymbol{C} - \boldsymbol{I})$ are readily obtained from equations (A2.7) and (A2.8).

Appendix 3

Legendre Transformation

Consider a function $\phi(x, y)$ of two variables, x and y, with differential

$$d\phi = \frac{\partial \phi}{\partial x} dx + \frac{\partial \phi}{\partial y} dy. \tag{A3.1}$$

Let

$$p = \frac{\partial \phi}{\partial x} \text{ and } q = \frac{\partial \phi}{\partial y}, \tag{A3.2}$$

so that equation (A3.1) becomes

$$d\phi = pdx + qdy. \tag{A3.3}$$

The purpose of the translation is to change the function ϕ of x and y to a function of the new variables p and q with its differential expressed in terms of dp and dq . Let $\psi(p, q)$ be defined by

$$\psi = \phi - px - qy, \tag{A3.4}$$

with differential,

$$d\psi = d\phi - pdx - xdp - qdy - ydq. \tag{A3.5}$$

Substituting equation (A3.3) in equation (A3.5) gives

$$d\psi = -xdp - ydq,$$

where

$$x = -\frac{\partial\psi}{\partial p}, \quad y = -\frac{\partial\psi}{\partial q}. \tag{A3.6}$$

Equation (A3.4) is a dual Legendre transformation. A single Legendre transformation involves one variable. For example, instead of equation (A3.4), let $\psi(p, y)$ be defined by

$$\psi = \phi - px, \tag{A3.7}$$

with differential

$$d\psi = d\phi - pdx - xdp. \tag{A3.8}$$

Substituting equation (A3.3) in equation (A3.8) gives

$$d\psi = qdy - xdp,$$

where

$$x = -\frac{\partial\psi}{\partial p}, \quad q = \frac{\partial\psi}{\partial y}. \tag{A3.9}$$

The potentials of thermoelasticity, which are fundamental equations of state, are the internal energy $u(\boldsymbol{C}, s)$, the Helmholtz free energy $f(\boldsymbol{C}, \Theta)$, the enthalpy $h(\boldsymbol{S}, s)$, and the Gibbs free energy $g(\boldsymbol{S}, \Theta)$. These are interrelated by Legendre transformations. An example of a single Legendre transformation in thermoelasticity theory is

$$f = u - \Theta s.$$

It follows from equation $(A3.9)_1$ that

$$s(\boldsymbol{C}, \Theta) = -\partial f/\partial\Theta, \text{ and } \boldsymbol{q}(\boldsymbol{C}, \Theta) = \partial f/\partial\boldsymbol{C}.$$

It can be shown otherwise that

$$\boldsymbol{q} = \boldsymbol{S}/(2\rho_0).$$

An example of a dual Legendre transformation in thermoelasticity theory is

$$g(\boldsymbol{S}, \Theta) = u - \frac{S_{KL}C_{KL}}{2\rho_0} - \Theta s.$$

It follows from equation (A3.6) that

$$s(\boldsymbol{S}, \Theta) = -\frac{\partial g}{\partial\Theta}, \; C_{KL}(\boldsymbol{S}, \Theta) = -2\rho_0\frac{\partial g}{\partial S_{KL}}.$$

Appendix 4

Linear Vector Spaces

In order to consider linear vector spaces it is useful to first consider a special case, the set of vectors of elementary vector analysis. The vector sum $\boldsymbol{x}+\boldsymbol{y}$ of any two vectors of the set is given by the parallelogram rule, and the set has the following properties:

a. $\boldsymbol{x}+\boldsymbol{y}=\boldsymbol{y}+\boldsymbol{x}$ (commutative property),
b. $\boldsymbol{x}+(\boldsymbol{y}+\boldsymbol{z})=(\boldsymbol{x}+\boldsymbol{y})+\boldsymbol{z}$ (associative property),
c. There exists a zero vector, $\mathbf{0}$, such that $\boldsymbol{x}+\mathbf{0}=\boldsymbol{x}$,
d. For every $\boldsymbol{x}$ there is a negative, $(-\boldsymbol{x})$, such that $\boldsymbol{x}+(-\boldsymbol{x})=\mathbf{0}$.

When a vector $\boldsymbol{x}$ is multiplied by a real scalar α the result is a vector $\alpha\boldsymbol{x}$, and scalar multiplication has the following properties:

e. $\alpha(\beta\boldsymbol{x})=\alpha\beta\boldsymbol{x}$ (associative property),
f. $(\alpha+\beta)\boldsymbol{x}=\alpha\boldsymbol{x}+\beta\boldsymbol{x}$ (distributive property),
g. $\alpha(\boldsymbol{x}+\boldsymbol{y})=\alpha\boldsymbol{x}+\alpha\boldsymbol{y}$ (distributive property).

The vector space of elementary vector analysis, denoted by V, is a special case of a Euclidean space since it has a scalar or inner product

$$\boldsymbol{x}\cdot\boldsymbol{y}=x_1y_1+x_2y_2+x_3y_3=x_iy_i, \tag{A4.1}$$

in terms of the components of vectors $\boldsymbol{x}$ and $\boldsymbol{y}$ referred to a three-dimensional rectangular Cartesian coordinate system. The scalar product has properties

h. $\boldsymbol{x}\cdot\boldsymbol{y}=\boldsymbol{y}\cdot\boldsymbol{x}$ (commutative property),
i. $\boldsymbol{x}\cdot(\boldsymbol{y}+\boldsymbol{z})=\boldsymbol{x}\cdot\boldsymbol{y}+\boldsymbol{x}\cdot\boldsymbol{z}$ (distributive property),
j. $\boldsymbol{x}\cdot\boldsymbol{x}\geqslant 0$ (equality holds when $\boldsymbol{x}=0$)

The relation (A4.1) implies that V is a three-dimensional space and the dimensionality of a vector space is made precise later in this appendix. A set of vectors that are normal to a particular vector, and includes the vector $\mathbf{0}$, is a two-dimensional subspace of V since properties (a) to (i) are valid with the scalar product given by $\boldsymbol{x} \cdot \boldsymbol{y} = x_1 y_1 + x_2 y_2$.

Now consider a set quantities $\boldsymbol{x}_1, \boldsymbol{x}_2, \boldsymbol{x}_3, \cdots \boldsymbol{x}_n$ that have properties analogous to (a) to (g) and have a scalar product $q(\boldsymbol{x}_i, \boldsymbol{x}_j)$ with properties analogous to (h) and (i). These set quantities constitute what is known as a Euclidean linear vector space or inner product space, and the elements are regarded as vectors in a general sense. That is, the space of vectors of elementary vector analysis is a special case. The vectors $\boldsymbol{x}_1, \boldsymbol{x}_2, \boldsymbol{x}_3, \cdots \boldsymbol{x}_n$ form a linearly independent system of order n if

$$\alpha_1 \boldsymbol{x}_1 + \alpha_2 \boldsymbol{x}_2 + \alpha_3 \boldsymbol{x}_3 + \cdots \alpha_n \boldsymbol{x}_n = 0$$

holds only when all the numbers $\alpha_1, \alpha_2, \alpha_3 \cdots \alpha_4$ are zero, otherwise the vectors are linearly dependent. The basis of a linear vector space is any linearly independent system maximum order. For example, for the three-dimensional space of vectors of elementary vector analysis any three non coplanar vectors $\boldsymbol{x}$, $\boldsymbol{y}$, and $\boldsymbol{z}$ is a basis. This means any vector $\boldsymbol{u}$ of the space can be expressed as a linear combination

$$\boldsymbol{u} = \alpha_1 \boldsymbol{x} + \alpha_2 \boldsymbol{y} + \alpha_3 \boldsymbol{z}$$

of the basis vectors.

In Cartesian tensor analysis the orthogonal system of unit base vectors $\{\mathbf{e}_1, \mathbf{e}_2, \mathbf{e}_3\}$ with $\mathbf{e}_i \cdot \mathbf{e}_j = \delta_{ij}$, is linearly independent, and a vector $\boldsymbol{u}$ can be expressed in rectangular Cartesian component form as

$$\boldsymbol{u} = u_1 \mathbf{e}_1 + u_2 \mathbf{e}_2 + u_3 \mathbf{e}_3$$

as shown chapter 1.

Another example of Euclidean linear vector space is the space of $(1 \times n)$ row matrices with vectors in the general sense given by vector $\mathbf{a} = (a_1, a_2, \cdots a_n)$. A basis for this space is the set

$$\{(1, 0, \cdots, 0), (0, 1, \cdots, 0), \cdots, (0, 0, \cdots, 1)\}$$

of n linearly independent vectors, and the scalar product of vectors $\mathbf{a}$ and $\mathbf{b}$ is given by

$$q(\mathbf{a}, \mathbf{b}) = \sum_{i=1}^{n} \mathrm{a}_i \mathrm{b}_i.$$

The space of vectors of elementary vector analysis is a special case with $n = 3$ (or $n = 2$ for a set of coplanar vectors) and each element of a row matrix regarded as a component of a vector in the sense of elementary vector analysis.

An important vector space is the set of second-order tensors, and this is shown to be a Euclidean linear vector space as follows. If $\boldsymbol{X}$, $\boldsymbol{Y}$, and $\boldsymbol{Z}$ are second-order tensors, referred to a three-dimensional point space, the properties (a) to (g), with $\boldsymbol{x}$, $\boldsymbol{y}$, and $\boldsymbol{z}$ replaced by $\boldsymbol{X}$, $\boldsymbol{Y}$, and $\boldsymbol{Z}$ are valid. Since a second-order tensor can be given in terms of Cartesian components by

$$\boldsymbol{X} = X_{ij} \mathbf{e}_i \otimes \mathbf{e}_j,$$

it follows that the space of second-order tensors is nine dimensional and $\mathbf{e}_i \otimes \mathbf{e}_j$ is a possible basis. This basis is orthogonal since the inner product of $\mathbf{e}_i \otimes \mathbf{e}_j$ and $\mathbf{e}_k \otimes \mathbf{e}_t$ is

$$\mathrm{tr}\left[(\mathbf{e}_i \otimes \mathbf{e}_j)^T (\mathbf{e}_k \otimes \mathbf{e}_t)\right] = \delta_{ik}\delta_{jt}.$$

A scalar product

$$q(\boldsymbol{X}, \boldsymbol{Y}) = \boldsymbol{X} \cdot \boldsymbol{Y} = \mathrm{tr}\left(\boldsymbol{X}\boldsymbol{Y}^T\right)$$

with $\boldsymbol{x}$, $\boldsymbol{y}$, and $\boldsymbol{z}$ replaced by $\boldsymbol{X}$, $\boldsymbol{Y}$, and $\boldsymbol{Z}$, has properties (h) to (j).

Index

ablation, 54
adiabatic deformation, 150
Almansi's strain tensor, 77, 147
alternating tensor, 26, 28
axial vector. *See* pseudo vector
azimuth angle, 255
azimuthal shear, 197

base vectors, 250
basic invariants, 21
basis vectors, 12
Biot strain tensor, 78
Blatz P.J., 169
body force per unit mass, 86
bound vector, 9
Boyle's law, 124
bulk modulus, 160, 209, 210
bulk viscosity, 125, 126, 221, 222, 239

canonical form, 23
Cartesian tensor analysis, 4
Cauchy elasticity, 153
Cauchy stress tensor. *See* stress tensors
Cauchy-Green tensors, 144
Cayley-Hamilton Theorem, 24, 158, 159
Chadwick P., 209, 216
change of entropy, 205
change of internal energy produced by deformation, 205
characteristic equation, 20, 22
Charles law, 124
Christensen R.M., 224, 240
classical elasticity, 126, 166
classically thermoelastic bar, 214
Clausius statement of the second law, 123, 208
Clausius-Duhem inequality, 118, 207
Clausius-Planck inequality, 120
coefficient of volume thermal expansion, 124, 213
cofactor, 29, 161
co-latitude angle, 255
column matrix, 12
commutative rule, 15
complementary energy, 148
compliance tensor, 128
compressible generalization of the Mooney-Rivlin, 169
compressible generalization of the neo-Hookean strain energy function, 169, 215
compressible isotropic models, 210
conjugate properties, 203
conjugate rate variable, 105
conjugate tensors, 159
conservation of energy, 115
conservation of mass, 63
constant entropy, 202
constitutive equation, 122
contact force, 84
continuous bodies, 50
contravariant, 4
coordinate curves, 250
coordinate surfaces, 250
covariant, 4
Creasy C.F.M., 209
creep, 224
creep function, 227, 247
current configuration, 90, 112
curvilinear coordinate system, 4, 250
cylindrical bar, 165

cylindrical polar coordinates, 175, 253, 259, 264
cylindrical polar form, 257
cylindrical symmetry, 165

d'Alembert's Principle, 114
deformation fields, 168
deformation gradient tensor, 58, 59, 70, 74, 108, 152
  dilatational part, 74
  isochoric part, 74
density, 86
  reference configuration, 99
  spatial configuration, 63, 99
determinant, 20, 28, 31, 58
determinism, 138
dilatation, 58, 210, 222
Dirac delta function, 229
displacement gradient tensor, 78
displacement vector, 76
dissipation inequality, 118
dissipation potential, 223
divergence, 42
divergence of a vector, 39
divergence operation, 100
divergence theorem, 44, 45, 100, 106, 112, 118
dyadic product, 11, 16

eigenvalues, 23, 24, 34, 35, 36, 91, 163
eigenvectors, 22, 23, 24, 34, 35, 36, 91 163
energetic, 205, 218
energetic response, 202
energy equation, 115, 122
enthalpy, 136
entropic, 205, 218
entropic response, 202
entropic thermoelasticity, 215
entropy, 118, 213
entropy inequality, 207
entropy inequality for a thermoelastic solid, 207
entropy jump across a shock, 208
entropy production, 208, 222
equation of equilibrium, 111
equation of motion, 106
equilibrium thermodynamics, 114
Ericksen J.L., 168
Eringen C., 138
Euclidean linear vector space, 271
Euclidean point space, 1
Euclidean transformation, 140
Euler's Equation, 60
Eulerian coordinates. *See* spatial coordinates
Eulerian strain tensor. *See* Almansi's strain tensor
Eulerian tensors, 90
eversion of a cylindrical tube, 185
experimental results, 212

finite axial strain, 214
finite deformation elastostatics for isotropic hyperelastic solids, 168
first law of thermodynamics, 115
first order tensor fields. *See* vector field
Flügge W., 224
foam rubbers, 160
Fourier's law, 115, 123
fourth order tensor, 26
frame of reference, 139
free suffix, 6
fundamental equation of state, 136, 203

Galilean frame of reference, 141
Galilean transformations, 143
generalized functions, 229
Gibbs free energy, 137
Gibbs relation, 132, 135, 204
Goodier, 127
gradient operator, 263
gravitational potential energy, 115
Green A.E., 180
Green deformation tensor, 156
Green elastic, 155
Green strain rate tensor, 156
Green's strain tensor, 75, 77, 78, 156

Hadamard strain energy function, 169
Haupt P., 114
heat conduction, 220
heat conduction tensor, 115, 123
heat flux vector, 115, 123, 138
Helmholtz free energy, 119, 131, 155, 202
hereditary integral, 233
Hill R., 104
Holzapfel G.A., 215
homogeneous deformation, 168

Hooke's law, 127
generalized, 127
Hunter S.C., 240
hydrostatic state of stress, 92, 95
hyperelastic solid, 155, 156, 160, 162
hyperelastic stress-deformation relations, 170
hyperelasticity, 157
hysteresis effects, 208, 220

ideal inviscid fluid, 123
ideal rubber, 212
impact modulus, 227, 232
improper orthogonal tensor, 32
incompressibility, 58, 160
incompressible fluid, 122
incompressible Hyperelastic solid, 160
incompressible isotropic models, 210
incompressible materials, 160
incompressible solid, 210
inertia tensor of the element, 69
inertial frame of reference, 143
infinitesimal rotation, 80
infinitesimal rotation tensor, 80
infinitesimal strain tensor, 78, 80, 160
inner product, 5, 17
internal energy, 124, 131, 155, 202, 212
internal variable, 231, 238
invariants, 158, 223
inverse, 30, 59
inverse methods, 168
inversion effect, 218
inviscid fluids, 220
isentropic deformation, 155
isentropic process, 120
isentropic simple tension, 213, 214, 216, 217
isentropic strain energy, 202
isochoric deformation, 160, 210
isochoric flow, 124, 125
isochoric motions, 58
isothermal, 150
isothermal deformation, 155, 213
isothermal moduli, 131
isothermal shear, 210
isothermal simple tension, 212, 213
isothermal strain energy, 202
isotropic elastic solid, 157
isotropic state of stress. *See* hydrostatic state of stress
isotropic tensor, 24, 28
isotropy, 153, 154, 160

Jacobian of the transformation, 58
Jaumann rate, 145
Joule heating, 116

Kelvin-Voigt element, 246
Kelvin-Voigt viscoelastic solid, 225, 228, 239
Kestin J., 202
Killing's theorem, 67
kinematically admissible virtual displacement field, 110
kinetic energy, 106, 155
Ko W.L., 169
Kronecker delta, 6, 17, 91

Lagrangian configuration. *See* reference configuration
Lagrangian multiplier, 93, 126, 160, 162, 180, 222
Lagrangian strain, 76
Lagrangian strain rate, 109
Lagrangian strain tensor, 204
Lamé's constants, 129
Laplace expansion, 28
Laplace transform, 236, 246
Laplacian operator, 261
laws of thermodynamics, 121
left Cauchy strain tensor, 75
left Cauchy-Green strain tensor, 172
left polar decompositions, 36
left stretch tensor, 72, 165
Legendre transformation, 131, 132, 135, 137, 203
Leibniz's rule, 62
Lighthill M.J., 230
linear theory of elasticity, 150
linear thermoelasticity, 130
linear vector spaces, 269
linear viscoelastic medium, 224
linearly independent system, 270
local action, 138
logarithmic strain tensors, 78
Love A.E.H., 127

Malvern L.E., 111
mass point, 50

material configuration. *See* reference configuration
material description, 50, 53, 77
material frame indifference, 138
material lines, 59
material objectivity, 167
material surface, 54, 59
material symmetry, 153
material volume, 59, 63
Mathematica, 24, 35, 37
matrix, 2
matrix multiplication, 1
Maxwell fluid, 225
mechanical compressibility, 210
mechanical dissipation, 243
mechanically incompressible fluid, 160
mechanically incompressible solid, 160
metric tensor, 252
modified deformation gradient tensor, 74
modified entropic elasticity, 218
Mooney-Rivlin material, 179
Mooney-Rivlin solid, 197
Mooney-Rivlin strain energy function, 166, 178
Müller I., 124
mutually orthogonal, 21, 22

natural reference configuration, 50
Navier-Stokes equation, 126
negative definite, 34
negative definite tensors, 36
neoHookean elastic solid, 167
neo-Hookean solid, 188
neo-Hookean strain energy function, 169
Newton's second law, 86
Newtonian viscous fluid, 125, 220, 239
nominal stress, 98, 112, 162
nominal stress tensor, 98
non-ideal rubbers, 209
non-newtonian viscous fluid, 223
non-singular second order tensor, 35
normal component, 85
normal stress, 89, 93
normalized eigenvectors, 21
numerical results, 217

objectivity, 151
oblique Cartesian coordinate system, 4
observer, 139, 140
octahedral plane, 96
octahedral shearing stress, 96
Ogden R.W., 102, 104, 174, 175, 189
Oldroyd J.G., 146
orthogonal curvilinear coordinate systems, 249, 258
orthogonal tensor, 32, 33, 36, 71
orthogonal transformations, 9, 15
orthogonality, 252
orthonormal, 6

parallelogram rule, 5
particle, 50
particle path, 52
partly energetic, 209
Pearson C.E., 95
perfectly elastic solid, 119
physical components, 253, 256
piezotropic model, 208
plane deformation, 171
Poisson's ratio, 169
polar decomposition theorem, 35, 70
polar media, 85
polar vector, 1, 10
polymers, 209
positive definite second order tensor, 21, 32, 33, 34, 35, 36
positive semidefinite, 33, 34
potential energy, 116
Poynting effect, 172, 179
principal axes, 34, 67
principal components, 164
principal directions, 91
principal invariants, 163, 173
principal stresses, 91, 92, 93
principal stretches, 71, 72, 163, 210
  modified stretches, 74
principle of local state, 114, 202, 243
principle of virtual work, 164
proper orthogonal tensor, 32, 71, 152, 154
pseudo scalar, 10
pseudo vector, 10
pure bending of an hyperelastic plate, 189
pure dilatation, 125, 130
pure hydrostatic stress, 95
pure shear stress, 94, 95

radiation, 116
rate of change of momentum of the body, 86

rate of deformation, 107, 113
rate of deformation tensor, 65, 67, 77, 159
rate of entropy production per unit mass, 208
rate of heat supply, 115, 123
rectangular Cartesian coordinate system, 1, 2, 3, 6
reference configuration, 50, 108, 112, 117
referential description, 50, 53
referential heat flux, 207
Reiner-Rivlin fluid, 223
relation between the velocity gradient and the deformation gradient tensor, 65
relative description, 51
relaxation function, 226, 232, 246
relaxation modulus, 224, 227, 246
relaxation time, 226, 237, 246
resultant moment acting on a body, 86
right Cauchy Green deformation tensor, 75, 152, 157, 204
right polar decompositions, 36
right stretch tensor, 72, 165
rigid body, 122
rigid body mechanics, 122
rigid body motion, 122
rigid motion, 121
Rivlin R.S., 189
rotation, 151
rate of deformation, 107, 113
rubber-like materials, 160
rubber-like solids, 210

scalar, 1, 2, 17
scalar differential operator, 40
scalar field, 38
scalar function, 223
scalar invariants, 9, 21
scalar product, 5, 32
Schlichting H., 222
second law of thermodynamics, 117, 124
second order tensor field, 38
shear and bulk relaxation functions, 234
shear deformation, 222
shear modulus, 160, 209
shear modulus for infinitesimal deformation, 169
shear viscosity, 221
shearing component, 85
shearing stress, 89
shearing stress on the element, 87
shock wave propagation, 220
shock waves, 208
simple materials, 138
simple shear, 125, 165, 171
simple shear deformation, 170
simple tension, 172, 212
simple tension of the neo-Hookean solid, 174
Sokolnikov I.S., 127, 128
spatial configuration, 51, 84, 90, 111, 113, 118, 154
spatial description, 51, 53
spatial form of continuity equation, 63
spatial heat flux, 207
specific enthalpy, 203
specific Gibbs free energy, 203
specific heat, 122, 133
specific heat at constant deformation, 209
specific Helmholtz free energy, 203
specific internal energy, 115, 203
specific strain energy, 156
spectral decomposition, 163
spectral form, 24, 34
spherical polar coordinates, 180, 254, 259, 264, 265
spherical polar form, 257
spherical symmetry, 165
spin tensor, 65, 70, 107, 148
spring-dashpot models, 224
square matrices, 13
square root of a second order tensor, 35, 36
simple shear, 125, 165, 171
spin tensor, 65, 70, 107, 148
standard material, 247
standard model, 230, 239
statically admissible stress, 112
statically admissible stress field, 110, 113
stationary values, 93
statistical theory of rubber elasticity, 212
steady flow, 52, 54
stiffness tensor, 127
Stokes' hypothesis, 126, 222
Stokes' theorem, 44
Stokesian fluid, 223
stored energy, 155

stored energy function, 155
strain energy, 239
strain energy function, 156, 157, 159
strain energy rate, 159
strain rate tensor, 78, 80
stream line, 52
stress boundary conditions, 162
stress discontinuities, 110
stress power, 155, 159, 221
stress relaxation, 224
stress tensor
  Biot, 101, 164, 214
  Cauchy, 2, 90, 91, 106, 150, 153, 158, 160, 164
    deviatoric part, 95
    isotropic part, 95
  first Piola-Kirchhoff, 107
  nominal, 107, 212
  second Piola-Kirchhoff, 100, 107, 153, 156, 158, 203
stress vector, 2, 84, 85
strictly entropic elasticity, 209, 212, 213, 215
substitution property, 17, 91
suffix notation, 4, 18
summation convention, 35
superposition principle, 227
surface traction, 85
symbolic form, 12
symbolic notation, 4, 17
symmetric part of the displacement gradient tensor, 80
symmetric second order tensor, 34
symmetry group, 154
symmetry transformation, 154

Tanner R.I., 220
telescopic shear of the neo-Hookean solid, 200
temperature, 121
tensor, 2
  first order. *See* vector
  fourth order, 25
  higher order, 2
  second order, 1, 12, 13, 34, 42
    antisymmetric part, 14
    deviatoric part, 18
    isotropic part, 18
    symmetric part, 13, 14, 22, 33, 34, 91
  third order, 25
tensor addition, 19
tensor character, 19
tensor contraction, 19
tensor field, 38
tensor multiplication, 18, 19
tensor product, 11, 25
tensor transformation rule, 18
thermal equation of state, 122
thermal strain, 130
thermal wave speed, 115
thermodynamic potentials, 202
thermodynamic pressure, 125, 221, 222
thermodynamic properties, 131
thermodynamics of irreversible processes, 114
thermoelastic inversion effect, 215
thermoelastic potentials, 131
third invariant, 29
third order isotropic tensor, 26
Thurston R.N., 240
Timoshenko S., 127
trace, 17
transformation matrix, 7, 15
transformation rules for base vectors, 8
transport equation, 63
transport properties, 126
transpose, 13
Tresca yield condition, 94, 98
triple scalar product, 28
true scalar, 10
Truesdell C.A., 168

unit base vectors, 7
unit step function, 226, 229

valanis-Landel hypothesis, 165
vector, 1, 3, 9, 12
vector field, 38, 41
vector gradient, 39, 41
vector magnitude, 9
velocity, 2, 54
velocity gradient tensor, 64
virtual displacements, 110
virtual work, 110
viscoelastic, 237
viscosity coefficient, 125, 238
viscous fluid, 124
Voigt matrix, 130

volume coefficient of thermal expansion, 160, 209, 210, 215, 216, 240
volume thermal expansion, 160
von Mises yield criterion, 97
vorticity vector, 67
White F.M., 126
Whitham, 220
Zerna W., 168, 180
zeroth order tensor field. *See* scalar field

Printed in the United States
By Bookmasters